PROBABILITY THEORY
WITH
THE ESSENTIAL ANALYSIS

APPLIED MATHEMATICS AND COMPUTATION

A Series of Graduate Textbooks, Monographs, Reference Works

Series Editor: ROBERT KALABA, University of Southern California

No. 1 MELVIN R. SCOTT
Invariant Imbedding and its Applications to Ordinary Differential Equations: An Introduction, 1973

No. 2 JOHN CASTI and ROBERT KALABA
Imbedding Methods in Applied Mathematics, 1973

No. 3 DONALD GREENSPAN
Discrete Models, 1973

No. 4 HARRIET H. KAGIWADA
System Identification: Methods and Applications, 1974

No. 5 V. K. MURPHY
The General Point Process: Applications to Bioscience, Medical Research, and Engineering, 1974

No. 6 HARRIET H. KAGIWADA and ROBERT KALABA
Integral Equations Via Imbedding Methods, 1974

No. 7 JULIUS T. TOU and RAFAEL C. GONZALEZ
Pattern Recognition Principles, 1974

No. 8 HARRIET H. KAGIWADA, ROBERT KALABA, and SUEO UENO
Multiple Scattering Processes: Inverse and Direct, 1975

No. 9 DONALD A. PIERRE and MICHAEL J. LOWE
Mathematical Programming Via Augmented Lagrangians: An Introduction with Computer Programs, 1975

No. 10 J. SUSAN MILTON and CHRIS P. TSOKOS
Probability Theory with the Essential Analysis, 1976

Other Numbers in preparation

PROBABILITY THEORY WITH THE ESSENTIAL ANALYSIS

J. SUSAN MILTON
Radford College
Radford, Virginia

CHRIS P. TSOKOS
University of South Florida
Tampa, Florida

1976
Addison-Wesley Publishing Company
Advanced Book Program
Reading, Massachusetts

London · Amsterdam · Don Mills, Ontario · Sydney · Tokyo

Library of Congress Cataloging in Publication Data

Milton, Janet Susan.
Probability theory with the essential analysis.

(Applied mathematics and computation ; no. 10)
Includes bibliographies and index.
1. Probabilities. 2. Axioms. 3. Measure theory.
I. Tsokos, Chris P., joint author. II. Title.
QA273.4.M54 519.2 76-27867
ISBN 0-201-07604-7
ISBN 0-201-07605-5 pbk.

Reproduced by Addison-Wesley Publishing Company, Inc., Advanced Book Program, Reading, Massachusetts, from camera-ready copy prepared by the authors.

American Mathematical Society (MOS) Subject Classification Scheme (1970): 60B05, 60B10, 60F05

Printed in the United States of America

ABCDEFGHIJ-HA-79876

To Our Parents

Enid and George Milton, Maria and Peter Tsokos

CONTENTS

SERIES EDITOR'S FOREWORD

Execution times of modern digital computers are measured in nanoseconds. They can solve hundreds of simultaneous ordinary differential equations with speed and accuracy. But what does this immense capability imply with regard to solving the scientific, engineering, economic, and social problems confronting mankind? Clearly, much effort has to be expended in finding answers to that question.

In some fields, it is not yet possible to write mathematical equations which accurately describe processes of interest. Here, the computer may be used simply to simulate a process and, perhaps, to observe the efficacy of different control processes. In others, a mathematical description may be available, but the equations are frequently difficult to solve numerically. In such cases, the difficulties may be faced squarely and possibly overcome; alternatively, formulations may be sought which are more compatible with the inherent capabilities of computers. Mathematics itself nourishes and is nourished by such developments.

Each order of magnitude increase in speed and memory size of computers requires a reexamination of computational techniques and an assessment of the new problems which may be brought within the realm of solution. Volumes in this series will provide indications of current thinking regarding problem formulations, mathematical analysis, and computational treatment.

ROBERT KALABA

PREFACE

During the past few years various problems arising in biology, chemistry, physics, engineering, and medicine which had previously been viewed from a deterministic viewpoint have been examined and re-examined in a probabilistic setting. There is now wide agreement that the stochastic approach to applied problems in the above areas is in many cases not only desirable but essential. The literature in these fields is becoming permeated with papers written with random models as their basis. For this reason it is necessary that students of the natural sciences as well as students of the mathematical sciences be thoroughly familiar with the basic language and tenets of probability theory.

We develop in this book the fundamental concepts of probability theory as an axiomatic system using a measure theoretic approach. However, no prior knowledge of measure and Lebesgue integration is assumed as the necessary material in these areas is presented in Chapters One and Two. These chapters are fairly rigorous in their approach and can be used as an introduction to measure and integration for the beginning student, as a review for the more advanced student or simply as a convenient reference for those more thoroughly versed in real analysis. We do assume a firm foundation in elementary calculus. We also assume a certain amount of mathematical maturity and experience in proof writing. The text should be readily accessible to advanced undergraduate (junior or senior level) and beginning graduate students in mathematics, statistics, operations research, chemistry, physics and engineering. We believe that this text can provide for these students a firm and mathematically correct foundation in probability theory which will enable them to continue in their study of the subject and to begin to grasp the implications of probability theory to their own disciplines.

This text can be effectively used in any of the following ways:

1. as a text for independent study courses or self study in analysis or probability theory;

2. a one quarter or one semester course in probability at an advanced level by utilizing Chapters Three through Seven;

3. as a one semester or two quarter course in measure, integration, and probability by utilizing Chapters Two through Seven;

4. as a two semester or three quarter course in analysis and probability at the junior or senior level by utilizing the entire book.

We believe that this text fills a basic current need, namely that for a readable and readily understandable introduction to probability theory for the beginning student. The material in the text has been organized in such a way that generally each theorem is provable from previous results contained within this book. Theorems for which this is not true which are felt to be beyond the scope of this text are indicated by * and references given. We firmly believe that mathematics must be learned by doing mathematics. The student is therefore urged to attempt to verify unstarred items for himself before consulting the proofs given in the text.

Items are numbered consecutively within each section of each chapter for easy reference. Each chapter is concluded with a summary of the main results obtained, a list of important terms and suggestions for further reading. We also supply for the convenience of the reader a list of special symbols together with the point at which they first appear in the text. Charts comparing measure theoretic and probabilistic terminology are given where applicable.

The text has been used successfully with both advanced undergraduate and graduate students.

We would like to acknowledge our students for their helpful discussions and suggestions during the preparation of the manuscript. A special thank you is due to Ms. Trudy Feldman for her excellent typing of the manuscript. In addition we would like to express our appreciation to Professor Robert Kalaba for his interest in and encouragement of this undertaking.

J. Susan Milton Chris P. Tokos

LIST OF SPECIAL SYMBOLS

Symbol	Meaning	First Reference
ε	is an element of	page 2
$\notin$	is not an element of	page 2
$\subseteq$	is a subset of	page 3
$=$	set equality	page 3
U	universal set	page 3
ϕ	empty set	page 3
N	natural numbers	page 4
$\cup$	set union	page 4
$\cap$	set intersection	page 4
$\times$	Cartesian Product	page 5
$X - A$	complement of A in X	page 6
A'	complement of A in U	page 6
$\overline{\lim}_{n\to\infty} A_n$	limit superior	page 11
$\underline{\lim}_{n\to\infty} A_n$	limit inferior	page 12
ρ_X	power set	page 14
τ	topology	page 16
(X,τ)	topological space	page 16
R	real numbers	page 19
E	Euclidean topology	page 19
τ_A	relative topology	page 20

Symbol	Meaning	First Reference
$f : X \to Y$	*function* f *maps* X *into* Y	*page 21*
$f^{-1}(A)$	*inverse image of* A *under* f	*page 21*
(x_n)	*sequence of real numbers*	*page 25*
$(x_n) \to x$	*sequence of real numbers* (x_n) *converges to the real number* x	*page 25*
(A_n)	*sequence of sets*	*page 12*
$\lim_{x \to x_0^+} f(x)=b$	*right hand limit*	*page 27*
$(x_n) \not\to x$	*sequence* (x_n) *of real numbers does not converge to* x	*page 25*
R_+	*positive real numbers*	*page 31*
(R^n, E^n)	*Euclidean n-space*	*page 33*
(X, C)	*measurable space*	*page 40*
β	*Borel sets of* R	*page 40*
$\overline{R}$	*extended real numbers*	*page 40*
λ	*measure*	*page 41*
(X, C, λ)	*measure space*	*page 42*
ℓ^*	*Lebesgue measure*	*page 43*
ℓ^*/β	*Lebesgue measure restricted to Borel sets*	*page 48*
β^n	*Borel sets of* R^n	*page 59*
$f = (f_1, f_2, \ldots, f_n)$	*vector of functions from* X *into* R^n	*page 60*
β^*	*Borel sets of* $(0, 1)$	*page 49*

Symbol	*Meaning*	*First Reference*
χ_A	*indicator function*	*page 63*
$M^+ = M^+(X, C)$	*non-negative measurable extended real valued functions*	*page 65*
$M = M(X, C)$	*measurable extended real valued functions*	*page 65*
$\int \psi d\lambda$	*Lebesgue integral*	*page 65*
inf S	*infimum of set* S	*page 73*
sup S	*supremum of set* S	*page 73*
$\int_E fd\lambda$	*Lebesgue integral of* f *over* E	*page 75*
$\limsup z_n$	*limit superior of sequence of extended real numbers*	*page 77*
$\liminf z_n$	*limit inferior of sequence of extended real numbers*	*page 77*
$\lim z_n$	*limit of sequence of real numbers*	*page 77*
(f_n)	*sequence of functions*	*page 78*
f_+	*positive part of* f	*page 84*
f_-	*negative part of* f	*page 84*
$S(Q{:}h;g)$	*Riemann-Stieltjes sum*	*page 90*
$\|\|Q\|\|$	*norm of a partition*	*page 91*
$\int_a^b hdg = \int_a^b h(x)dg(x)$	*Riemann-Stieltjes integral*	*page 91*
$\int_a^\infty hdg$	*infinite integral*	*page 91*
$\binom{n}{k}$	*combinatorial notation*	*page 92*

Symbol	Meaning	First Reference
n!	factorial notation	page 92
$\Gamma(z)$	gamma function	page 94
$(\Omega, \mathfrak{f}, P)$	probability space	page 100
P	probability measure	page 100
Ω	sample space	page 101
$P[E|F]$	conditional probability	page 125
X	random variable	page 141
$f \circ X$	composite map	page 147
$f(X)$	composite map	page 147
[ ]	greatest integer function	page 148
F	distribution function for random variable X	page 150
f	density function for random variable X	page 155
(R, β, P')	induced space	page 162
μ_g	Lebesgue-Stieltjes measure	page 163
$\int dF_X$	Lebesgue-Stieltjes integral	page 167
$\int dF$	Riemann-Stieltjes integral	page 167
$E[X]$	expected value of X	page 171
P.a.e.	P-almost everywhere	page 175
μ_n	nth ordinary moment	page 192
μ	mean	page 193
η_n	nth central moment	page 193

Symbol	*Meaning*	*First Reference*
σ^2	*variance*	*page 193*
$\mathrm{Var}\ X$	*variance of* X	*page 193*
σ	*standard deviation*	*page 193*
$m_X(\theta)$	*moment generating function*	*page 194*
$\dfrac{d^n m_X(\theta)}{d\theta^n}$	nth *derivative of* $m_X(\theta)$	*page 198*
$(X_n) \to X$	*convergence everywhere on* A	*page 210*
$(X_n) \underset{P.a.e.}{\to} X$	*convergence with probability one*	*page 214*
$(X_n) \underset{a.e.}{\to} X$	*convergence with probability one*	*page 214*
$(X_n) \not\to X$	*sequence does not converge to* X	*page 216*
$(X_n) \xrightarrow{P} X$	*convergence in probability*	*page 221*
$(X_n) \underset{m.s.}{\to} X$	*convergence in mean square*	*page 222*
$(X_1, X_2, \ldots, X_n)$	*random vector*	*page 234*
$H_{X_{\gamma_1}, X_{\gamma_2}, \ldots, X_{\gamma_n}}$	*joint distribution function for* $(X_{\gamma_1}\ X_{\gamma_2}, \ldots\ X_{\gamma_n})$	*page 255*

PROBABILITY THEORY
WITH
THE ESSENTIAL ANALYSIS

CHAPTER ONE

SET THEORY AND SOME TOPOLOGICAL ASPECTS OF EUCLIDEAN TOPOLOGY ON THE REAL LINE

1.0 *INTRODUCTION*

The purpose of this Chapter and Chapter Two is to review the basic mathematics which will be necessary in our approach to the study of *probability theory*. The reader with a firm background in *set theory*, *measure theory* and *integration*, and *analysis* can skip these chapters and proceed to Chapter Three. Throughout this book we shall, however, refer to various results obtained here. In many instances in this chapter we shall simply state pertinent results either leaving the verification to the reader in the more straightforward instances or else referencing a source where a detailed proof may be found. If a source is not indicated then it is felt that the result is easily provable from preceeding material.

1.1 *SET THEORY*

By a *set* we shall mean any collection of objects.

We shall follow the usual convention of denoting sets by upper case letters A, B, X, Ω and so forth. We shall use the term "set" and "collection" interchangeably. We shall also be using the standard set builder notation. That is, sets X will occasionally be defined using a listing process or a simplified descriptive process of the form $X = \{x : Q(x)\}$ where by the latter notation we mean to say "X is the set of all objects x such that the statement Q(x) is true."

Example 1.1.1

The set M of all even numbers can be described by either listing or by use of a descriptive process as follows:

$$M = \{ . . . , -4, -2, 0, 2, 4, 6, . . . \}$$

or

$$M = \{n : n = 2k, \text{ for } k \text{ an integer}\}.$$

Let X be a set and let x be an object in X. We say that x is an element of set X or x is a member of set X. We shall write $x \in X$. If x is not a member of X, we write $x \notin X$.

Example 1.1.2

Referring to the set M of *Example* (1.1.1), $2 \in M$.

Definition 1.1.3

Let X be a set. A set A is said to be a *subset* of X (or A is contained in X), denoted $A \subseteq X$, if and only if every element of A is also an element of X.

Example 1.1.4

Referring to the set M of *Example* (1.1.1) the set A given by

$$A = \{n : n = 2k, \quad n \text{ a positive integer}\}$$

is a subset of M.

Example 1.1.5

$X \subseteq X$ for any set X.

Definition 1.1.6

Let A and B be sets. We say that A and B are *equal*, denoted $A = B$, if and only if $A \subseteq B$ and $B \subseteq A$.

Definition 1.1.7

The set of all objects under discussion in a given situation is referred to as the *universal set*.

Note that in most instances it is clear from the context of the discussion which set can be thought of as the universal set. If not, or if we wish to emphasize the universal set, we shall denote it specifically by $\mathcal{U}$.

Definition 1.1.8

The set with no elements is called the *empty* set or the *null* set.

We shall denote the null set by ϕ.

Example 1.1.9

$\phi \subseteq X$ for any set X.

Definition 1.1.10

A set X is said to be *countable* if it is either finite or its elements can be put into a one to one correspondence with the set N of natural numbers, where by *natural numbers* we shall mean $\{1, 2, 3, \ldots\}$. If X is not countable it is said to be uncountable.

Definition 1.1.11

Let $\{A_\gamma : \gamma \varepsilon \Gamma\}$ be a collection of sets. By the *union* of this collection denoted

$$\cup\{A_\gamma : \gamma \varepsilon \Gamma\} = \bigcup_{\gamma \varepsilon \Gamma} A_\gamma$$

we shall mean

$$\{x : x \varepsilon A_\gamma \quad \text{for some} \quad \gamma \varepsilon \Gamma\}.$$

Definition 1.1.12

Let $\{A_\gamma : \gamma \varepsilon \Gamma\}$ be a collection of sets. By the *intersection* of this collection denoted

$$\cap\{A_\gamma : \gamma \varepsilon \Gamma\} = \bigcap_{\gamma \varepsilon \Gamma} A_\gamma$$

we shall mean

$$\{x : x \varepsilon A_\gamma \quad \text{for each} \quad \gamma \varepsilon \Gamma\}.$$

Note that the index set Γ is essentially unspecified. It can be either countable or uncountable.

The following facts concerning *countability* will be of use later. We state these results now for convenience.

Theorem 1.1.13 (*Properties of Countable Sets*)

i) Any subset of a countable set is countable.

ii) If $A \subseteq X$ and A is uncountable, then X is uncountable.

iii) The set $N \times N$ of all ordered pairs of natural numbers is countable.

iv) The union of a countable collection of countable sets is countable.

v) The intersection of any collection of countable sets is countable.

Hint: Property i) is obvious and property ii) is its contrapositive. Property iii) can be proved by considering the array:

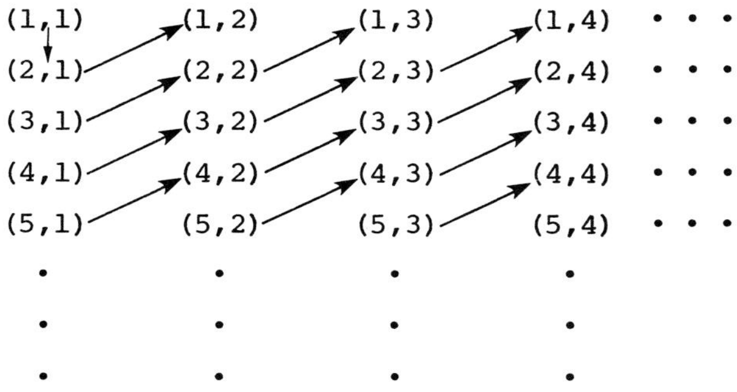

and using the indicated diagonalization procedure to define a one to one map from N onto $N \times N$. Property iv) follows from property iii) and the fact that if there exists a one to one map from a countable set X

onto a set S, then S is also countable. Property v) follows from property i) and the fact that the intersection of any collection of sets is a subset of each of the sets in the collection.

Definition 1.1.14

Let X and A be sets. The complement of A in X, denoted X - A, is defined to be

$$\{x : x \in X, \quad x \notin A\}.$$

Note 1.1.15

If we let $X = \mathcal{U}$ the universal set, then $X - A = \mathcal{U} - A = \{x : x \in \mathcal{U}, \quad x \notin A\}$. In this case the set $\mathcal{U} - A$ is usually referred to as simply the complement of A and is denoted by A'.

Note 1.1.16

Let $A \subseteq \mathcal{U}$, then $A \cup A' = \mathcal{U}$ and $A \cap A' = \phi$.

Theorem 1.1.17

Let X be a set and $A \subseteq X$. Then $X - [X - A] = A$.

Note 1.1.18

If $A \subseteq \mathcal{U}$ and $B \subseteq \mathcal{U}$ then $B - A = B \cap A'$.

Definition 1.1.19

Let A and B be sets. A and B are said to be *disjoint* if $A \cap B = \phi$.

Theorem 1.1.20

De Morgan's Laws Let X be a set and $\{A_\gamma : \gamma \in \Gamma\}$ be a collection of subsets of X. Then

i) $X - \bigcup_{\gamma \varepsilon \Gamma} A_\gamma = \bigcap_{\gamma \varepsilon \Gamma} [X - A_\gamma]$;

and

ii) $X - \bigcap_{\gamma \varepsilon \Gamma} A_\gamma = \bigcup_{\gamma \varepsilon \Gamma} [X - A_\gamma]$.

Proof: i) Let $x \varepsilon X - \bigcup_{\gamma \varepsilon \Gamma} A_\gamma$. Then $x \varepsilon X$ but $x \notin \bigcup_{\gamma \varepsilon \Gamma} A_\gamma$ by *Definition* (1.1.14). By *Definition* (1.1.11) $x \notin A_\gamma$ for any $\gamma \varepsilon \Gamma$. Thus $x \varepsilon X - A_\gamma$ for each $\gamma \varepsilon \Gamma$. By *Definition* (1.1.12) $x \varepsilon \bigcap_{\gamma \varepsilon \Gamma} [X - A_\gamma]$. This implies that $X - \bigcup_{\gamma \varepsilon \Gamma} A_\gamma \subseteq \bigcap_{\gamma \varepsilon \Gamma} [X - A_\gamma]$. Now let $x \varepsilon \bigcap_{\gamma \varepsilon \Gamma} [X - A_\gamma]$. Thus $x \varepsilon X - A_\gamma$ for each γ by *Definition* (1.1.12) and hence $x \varepsilon X$ but $x \notin A_\gamma$ for any $\gamma \varepsilon \Gamma$. By *Definition* (1.1.11) $x \notin \bigcup_{\gamma \varepsilon \Gamma} A_\gamma$. Hence $x \varepsilon X - \bigcup_{\gamma \varepsilon \Gamma} A_\gamma$. We have by *Definition* (1.1.3) that $\bigcap_{\gamma \varepsilon \Gamma} [X - A_\gamma] \subseteq X - \bigcup_{\gamma \varepsilon \Gamma} A_\gamma$. Applying *Definition* (1.1.6) we obtain that $X - \bigcup_{\gamma \varepsilon \Gamma} A_\gamma = \bigcap_{\gamma \varepsilon \Gamma} [X - A_\gamma]$.

To prove ii) note that

$$
\begin{aligned}
X - \bigcap_{\gamma \varepsilon \Gamma} A_\gamma &= X - \bigcap_{\gamma \varepsilon \Gamma} [X - (X - A_\gamma)] && \text{by } \textit{Theorem } (1.1.17) \\
&= X - (X - \bigcup_{\gamma \varepsilon \Gamma} [X - A_\gamma]) && \text{by part i)} \\
&= \bigcup_{\gamma \varepsilon \Gamma} [X - A_\gamma] && \text{by } \textit{Theorem } (1.1.17).
\end{aligned}
$$

When X is the universal set $\mathcal{U}$, De Morgan's Laws take on the following, and perhaps more familiar form.

Corollary 1.1.21

Let $\mathcal{U}$ be the universal set and $\{A_\gamma : \gamma \varepsilon \Gamma\}$ a collection of subsets of $\mathcal{U}$. Then

i) $\left(\bigcup_{\gamma\varepsilon\Gamma} A_\gamma\right)' = \bigcap_{\gamma\varepsilon\Gamma} A'_\gamma$;

ii) $\left(\bigcap_{\gamma\varepsilon\Gamma} A_\gamma\right)' = \bigcup_{\gamma\varepsilon\Gamma} A_\gamma{}'$.

Definition 1.1.22

Let X be a set and let $\mathcal{C} = \{A_\gamma : \gamma \varepsilon \Gamma\}$ be a collection of subsets of X. $\mathcal{C}$ is said to be a σ *algebra* (σ *field or Borel field*) if the following properties hold:

i) $\phi \varepsilon \mathcal{C}$;

ii) If $A \varepsilon \mathcal{C}$, then $X - A \varepsilon \mathcal{C}$;

iii) The union of any countable collection of elements of $\mathcal{C}$ is an element of $\mathcal{C}$.

Note 1.1.23

If it is assumed that property ii) holds, then it is evident that $\phi \varepsilon \mathcal{C}$ if and only if $X \varepsilon \mathcal{C}$. Similarly, by making use of property ii) and De Morgan's Laws it can be shown that the union of any countable collection of elements of $\mathcal{C}$ is an element of $\mathcal{C}$ if and only if the intersection of this collection is also an element of $\mathcal{C}$. These facts can be utilized to obtain an equivalent definition for the term σ algebra by replacing property i) with the statement "$X \in \mathcal{C}$" and property iii) with the statement "the intersection of any countable collection of elements of $\mathcal{C}$ is an element of $\mathcal{C}$."

Example 1.1.24

The following are examples of σ algebras:

i) Let X be any set and let $\mathcal{C}$ be the set of all subsets of X;

ii) Let X be a set, $X \neq \phi$. Let $\mathcal{C} = \{X, \phi\}$;

iii) Let X be the set of all real numbers and let $\mathcal{C}$ be the collection of all countable subsets of X, together with their complements with respect to X.

Verification

Parts i) and ii) are obvious. For part iii) we may assume without loss of generality that $X = \mathcal{U}$ so that the complement of a set A with respect to X may be expressed simply as A'. Since ϕ is countable, condition i) of *Definition* (1.1.22) is easily satisfied. By definition of $\mathcal{C}$, the complement of any element of $\mathcal{C}$ is an element of $\mathcal{C}$ and condition ii) is satisfied. Let $\{A_1, A_2, A_3, \ldots\}$ be any countable collection of elements of $\mathcal{C}$. Either A_i is countable for all i or A_i is uncountable for all i or there exists a subcollection $\{A_\gamma : \gamma \in \Gamma, \Gamma \subseteq N\}$ such that for $\gamma \in \Gamma$, A_γ is uncountable and for $\gamma \in N - \Gamma$, A_γ is countable. In the first case, *Theorem* (1.1.13), part iv), insures that

$\bigcup_{i=1}^{\infty} A_i$ is countable and therefore an element of $\mathcal{C}$. In the second case, for all i A_i is the complement of a countable set B_i. Hence we can write

$$\bigcup_{i=1}^{\infty} A_i = \bigcup_{i=1}^{\infty} B_i' = [\bigcap_{i=1}^{\infty} B_i]'$$

by use of *Corollary* (1.1.21). *Theorem* (1.1.13), part v), implies that $\bigcap_{i=1}^{\infty} B_i$ is countable and hence we have expressed $\bigcup_{i=1}^{\infty} A_i$ as the complement of a countable set. In the third instance note that

$$\bigcup_{i=1}^{\infty} A_i = [\bigcup_{\gamma \varepsilon \Gamma} A_\gamma] \cup [\bigcup_{\gamma \varepsilon N-\Gamma} A_\gamma].$$

For each $\gamma \varepsilon \Gamma$ we can express $A_\gamma = B_\gamma'$ where B_γ is countable. Hence using De Morgan's Laws we can obtain

$$\begin{aligned}\bigcup_{i=1}^{\infty} A_i &= [\bigcup_{\gamma \varepsilon \Gamma} A_\gamma] \cup [\bigcup_{\gamma \varepsilon N-\Gamma} A_\gamma] \\ &= [\bigcup_{\gamma \varepsilon \Gamma} B_\gamma'] \cup [\bigcup_{\gamma \varepsilon N-\Gamma} A_\gamma] \\ &= [\bigcap_{\gamma \varepsilon \Gamma} B_\gamma]' \cup [[\bigcup_{\gamma \varepsilon N-\Gamma} A_\gamma]']' \\ &= [[\bigcap_{\gamma \varepsilon \Gamma} B_\gamma] \cap [\bigcup_{\gamma \varepsilon N-\Gamma} A_\gamma]']'\end{aligned}$$

Note that by *Theorem* (1.1.13), part v) $\bigcap_{\gamma\varepsilon\Gamma} B_\gamma$ is countable and that

$$[\bigcap_{\gamma\varepsilon\Gamma} B_\gamma] \cap [\bigcup_{\gamma\varepsilon N-\Gamma} A_\gamma]' \subseteq [\bigcap_{\gamma\varepsilon\Gamma} B_\gamma].$$

Thus

$$[\bigcap_{\gamma\varepsilon\Gamma} B_\gamma] \cap [\bigcup_{\gamma\varepsilon N-\Gamma} A_\gamma]'$$

is countable and once again we have been able to express $\bigcup_{i=1}^{\infty} A_i$ as the complement of a countable set.

Definition 1.1.25

Let X be a set and let $\mathcal{B}$ be a non-empty collection of subsets of X. Let $\mathcal{C}$ be a σ algebra of subsets of X. $\mathcal{C}$ is said to be the *smallest* σ *algebra containing* $\mathcal{B}$ or the σ *algebra generated by* $\mathcal{B}$ if $\mathcal{C}$ is such that if $\mathcal{K}$ is any σ algebra containing $\mathcal{B}$ then $\mathcal{C} \subseteq \mathcal{K}$.

Theorem 1.1.26

Let X be a set and $\mathcal{B}$ a non-empty collection of subsets of X. There exists a smallest σ algebra of subsets of X containing $\mathcal{B}$.

Hint: Consider the intersection of all σ algebras of subsets of X containing $\mathcal{B}$ and argue that this intersection is a σ algebra.

Definition 1.1.27

Let $\mathcal{C} = \{A_n : n = 1, 2, 3, 4, \ldots\}$ be a sequence of sets. The *limit superior* of $\mathcal{C}$, denoted $\limsup_{n\to\infty} A_n$

or $\overline{\lim_{n\to\infty}} A_n$, is defined by

$$\overline{\lim_{n\to\infty}} A_n = \{x : x \in A_n \text{ for infinitely many values of } n\}.$$

Definition 1.1.28

Let $C = \{A_n : n = 1, 2, 3, \ldots\}$ be a sequence of sets. We define the *limit inferior* of C, denoted

$\lim_{n\to\infty} \inf A_n$ or $\underline{\lim}_{n\to\infty} A_n$, by

$$\underline{\lim}_{n\to\infty} A_n = \{x : x \in A_n \text{ for all but finitely many } n\}.$$

Example 1.1.29

Let $U = N$ where N denotes the natural numbers. Define a sequence of sets (A_n) as follows:

$A_1 = N - \{1\}$

$A_2 = N - \{2\}$

$A_3 = N - \{1,3\}$

$A_4 = N - \{2,4\}$

$A_5 = N - \{1,3,5\}$

.

.

.

$\overline{\lim_{n\to\infty}} A_n = N$

$\underline{\lim}_{n\to\infty} A_n = \phi$

Theorem 1.1.30

Let $C = \{A_n : n = 1, 2, 3, 4, \ldots\}$ be a sequence of sets. Then

$$\underline{\lim}_{n\to\infty} A_n \subseteq \overline{\lim}_{n\to\infty} A_n .$$

Note 1.1.31

The converse of *Theorem* (1.1.30) is not true in general as seen by *Example* (1.1.29).

Theorem 1.1.32

Let $C = \{A_n : n = 1, 2, 3, \ldots\}$ be a sequence of sets. Then

i) $$\overline{\lim}_{n\to\infty} A_n = \bigcap_{n=1}^{\infty} \bigcup_{m=n}^{\infty} A_m$$

and

ii) $$\underline{\lim}_{n\to\infty} A_n = \bigcup_{n=1}^{\infty} \bigcap_{m=n}^{\infty} A_m .$$

Proof: i) We shall use an indirect method of proof. Assume that $x \notin \bigcap_{n=1}^{\infty} \bigcup_{m=n}^{\infty} A_m$. Then there exists a natural number k such that $x \notin \bigcup_{m=k}^{\infty} A_m$. Hence $x \in A_m$ for only a finite number of indices n. Thus $x \notin \overline{\lim}_{n\to\infty} A_n$. This is sufficient to show that $\overline{\lim}_{n\to\infty} A_n \subseteq \bigcap_{n=1}^{\infty} \bigcup_{m=n}^{\infty} A_m$. Now assume that $x \notin \overline{\lim}_{n\to\infty} A_n$. Thus $x \in A_n$ for only a finite number of indices n. This implies that there exists a natural number k such that for $m \geq k$, $x \notin A_m$. Hence $x \notin \bigcup_{m=k}^{\infty} A_m$. This in turn implies that $x \notin \bigcap_{n=1}^{\infty} \bigcup_{m=n}^{\infty} A_m$.

This is sufficient to show that $\bigcap_{n=1}^{\infty} \bigcup_{m=n}^{\infty} A_m \subseteq \overline{\lim_{n\to\infty}} A_n$.

Thus by *Definition* (1.1.6) $\overline{\lim_{n\to\infty}} A_n = \bigcap_{n=1}^{\infty} \bigcup_{m=n}^{\infty} A_m$.

ii) Assume that $x \notin \bigcup_{n=1}^{\infty} \bigcap_{m=n}^{\infty} A_m$. By definition of the term set union $x \notin \bigcap_{m=n}^{\infty} A_m$ for any n. Thus given any natural number k there exists a natural number $\ell > k$ such that $x \notin A_\ell$. Thus $x \notin A_m$ for infinitely many subscripts m and therefore $x \notin \underline{\lim_{n\to\infty}} A_n$. This is sufficient to show that $\underline{\lim_{n\to\infty}} A_n \subseteq \bigcup_{n=1}^{\infty} \bigcap_{m=n}^{\infty} A_m$. Now assume that $x \notin \underline{\lim_{n\to\infty}} A_n$. This implies that $x \notin A_m$ for infinitely many subscripts m. Hence given any natural number n there exists a natural number $\ell > n$ such that $x \notin A_\ell$. Thus $x \notin \bigcap_{m=n}^{\infty} A_m$ for any n and therefore $x \notin \bigcup_{n=1}^{\infty} \bigcap_{m=n}^{\infty} A_m$.

This is sufficient to show that $\bigcup_{n=1}^{\infty} \bigcap_{m=n}^{\infty} A_m \subseteq \underline{\lim_{n\to\infty}} A_n$.

By *Definition* (1.1.6) $\underline{\lim_{n\to\infty}} A_n = \bigcup_{n=1}^{\infty} \bigcap_{m=n}^{\infty} A_m$.

Definition 1.1.33

Let X be a set and let ρ_X be the collection of all subsets of X. ρ_X is called the *power set* of X.

Definition 1.1.34

Let X be a set. A *partition* of X is a collection $\{E_\gamma : \gamma \in \Gamma\}$ of subsets of X such that

$\bigcup_{\gamma \in \Gamma} E_\gamma = X$ and $E_\alpha \cap E_\beta = \phi$, $\alpha \neq \beta$.

1.1 *EXERCISES*

1. Prove the general distributive laws for set unions and intersections. That is, show that

   i) $A \cap (\bigcup_{\gamma \in \Gamma} B_\gamma) = \bigcup_{\gamma \in \Gamma} (A \cap B_\gamma)$;

   and

   ii) $A \cup (\bigcap_{\gamma \in \Gamma} B_\gamma) = \bigcap_{\gamma \in \Gamma} (A \cup B_\gamma)$.

2. Prove that

   i) $A'' = A$;

   ii) $(A' \cap B')' = A \cup B$;

   iii) $(A' \cup B')' = A \cap B$;

   iv) $A \subseteq B$ if and only if $B' \subseteq A'$;

3. Show that

   i) $(\overline{\lim_{n\to\infty}} A_n)' = \underline{\lim_{n\to\infty}} (A_n)'$;

   ii) What is $(\underline{\lim_{n\to\infty}} A_n)'$? Verify your answer.

4. Let X be a non-empty set and let $\mathcal{C} = \{A_\gamma : A_\gamma \subseteq X, \gamma \in \Gamma\}$ be a σ algebra. Consider any sequence $A_{\gamma_1}, A_{\gamma_2}, A_{\gamma_3}, \ldots$ of elements of $\mathcal{C}$.

   Prove that $\overline{\lim_{n\to\infty}} A_{\gamma_n}$ and $\underline{\lim_{n\to\infty}} A_{\gamma_n}$ are elements of $\mathcal{C}$.

5. Suppose that $A_1 = A_3 = A_5 = \cdots = A$ and $A_2 = A_4 = A_6 = \cdots = B$. Find

$\overline{\lim_{n\to\infty}} A_n$ and $\underline{\lim_{n\to\infty}} A_n$.

6. Let $A_1, A_2, A_3, \ldots$ be a sequence of non-empty sets such that $A_n \subseteq A_{n+1}$ for each n. Find $\overline{\lim_{n\to\infty}} A_n$ and $\underline{\lim_{n\to\infty}} A_n$. Give an example of such a sequence.

7. Let $A_1, A_2, A_3, \ldots$ be a sequence of non-empty sets such that $A_{n+1} \subseteq A_n$ for each n. Find $\overline{\lim_{n\to\infty}} A_n$ and $\underline{\lim_{n\to\infty}} A_n$. Give an example of such a sequence.

1.2 *SOME TOPOLOGICAL ASPECTS OF EUCLIDEAN TOPOLOGY ON THE REAL LINE - EUCLIDEAN n-SPACE*

We shall point out in this section some of the pertinent properties of the Euclidean topology on the real line and indicate how Euclidean n-space can be defined. For a more detailed discussion of these topics the reader is referred to Greever [2].

Definition 1.2.1

A *topological space* is an ordered pair (X, τ) where X is a set and τ is a collection of subsets of X such that

i) $\phi \in \tau$ and $X \in \tau$;

ii) $U \cap V \in \tau$ whenever U and $V \in \tau$;

iii) $\bigcup_{\gamma\in\Gamma} U_\gamma \in \tau$ whenever $\{U_\gamma : \gamma \in \Gamma\} \subseteq \tau$.

τ is called a topology for X and the elements of τ are called open sets.

Examples 1.2.2

i) Let X be a set and let $\tau = \rho_X$. (See *Definition* (1.1.33)) The ordered pair (X, τ) is a topological space. The topology τ is sometimes referred to as the discrete topology for X.

ii) Let $X = \{a, b, c, d\}$ and let $\tau = \{\{a\}. \{a, b\}, \{a, c\}, \{a, b, c\}, X, \phi\}$. The ordered pair (X, τ) is a topological space.

iii) Let X be the set of all real numbers. Let $\tau = \{\phi\} \cup \{X - C : C \text{ is a countable subset of } X\}$. The ordered pair (X, τ) is a topological space.

Proof: i) and ii) are obvious. To see that example iii) is a topological space, note first that $\phi \in \tau$ by definition and that $X \in \tau$ since $X = X - \phi$ and ϕ is countable. Let $U = X - C_1$ and $V = X - C_2 \in \tau$. Then $U \cap V = (X - C_1) \cap (X - C_2) = X - (C_1 \cup C_2)$ by *Theorem* (1.1.20). Since C_1 and C_2 are countable, $C_1 \cup C_2$ is countable by *Theorem* (1.1.13). Hence $U \cap V$ has been expressed in the required form. Let $\{U_\gamma : \gamma \in \Gamma\}$ be any collection of elements of τ. Then for each γ, $U_\gamma = X - C_\gamma$ for some countable subset of X.

$\bigcup_{\gamma\in\Gamma} U_\gamma = \bigcup_{\gamma\in\Gamma} (X - C_\gamma) = X - \bigcap_{\gamma\in\Gamma} C_\gamma$ by *Theorem* (1.1.20).

However $\bigcap_{\gamma\varepsilon\Gamma} C_\gamma$ is countable by *Theorem* (1.1.13). Thus $\bigcup_{\gamma\varepsilon\Gamma} U_\gamma$ has been expressed in the desired form.

Definition 1.2.3

Let (X, τ) be a topological space and let $\mathcal{B}$ be a collection of subsets of X. $\mathcal{B}$ is called a *base* for the topology τ if and only if τ consists of ϕ and those sets which are unions of sets in $\mathcal{B}$. Elements of $\mathcal{B}$ are often referred to as *basic open sets*.

It is of interest to note that due to *Definition* (1.2.3), every open set U, that is every element $U \varepsilon \tau$, has the property that U contains a basic open set B about each of its points.

It is often convenient to start with a particular collection $\mathcal{B}$ of subsets of some set X and try to use this collection to generate a topology on X. It is evident that not every collection of subsets of X can serve as a base for a topology. Thus it is important to obtain a characterization for those collections which will generate a topology.

Examples 1.2.4

i) Let $X = N$ and $\tau = \rho_X$. Let $\mathcal{B} = \{\{1\}, \{2\}, \{3\}, \{4\}, . . .\}$. As seen in *Example* (1.2.2), (X, τ) is a topological space. $\mathcal{B}$ is a base for the topology τ.

ii) Let $X = \{a, b, c, d\}$. Let $\mathcal{B} = \{\{a\}, \{b\}, \{c\}\}$. $\mathcal{B}$ is not a base for any topology on X.

The desired characterization is obtained by the following theorem whose proof may be found in Greever [2].

Theorem 1.2.5*

Let X be a set and let $\mathcal{B}$ be a collection of subsets of X. Then there exists a unique topology τ for X such that $\mathcal{B}$ is a base for τ if and only if

i) for each $x \in X$, there exists a $B \in \mathcal{B}$ such that $x \in B$;

ii) given B_1 and $B_2 \in \mathcal{B}$ and $x \in B_1 \cap B_2$ there exists a $B \in \mathcal{B}$ such that $x \in B \subseteq B_1 \cap B_2$.

The technique generally employed to define what is called the "usual" or Euclidean topology for the set $\mathcal{R}$ of real numbers is to define a particular collection, $\mathcal{B}$, of subsets of $\mathcal{R}$ which satisfy conditions i) and ii) of *Theorem* (1.2.5). This collection can thus be used as a base for a unique topology on $\mathcal{R}$.

Definition 1.2.6

Let $\mathcal{R}$ be the set of real numbers and for each element $p \in \mathcal{R}$ and each real number $\varepsilon > 0$, define $N_\varepsilon(p) = \{x : x \in \mathcal{R}, |x - p| < \varepsilon\}$. Let $\mathcal{B} = \{N_\varepsilon(p) : p \in \mathcal{R}, \varepsilon > 0\}$. Let $\mathcal{E}$ be the unique topology on $\mathcal{R}$ with base $\mathcal{B}$. $\mathcal{E}$ is called the usual or Euclidean topology for $\mathcal{R}$.

It is conceivable that two different collections $\mathcal{B}_1$ and $\mathcal{B}_2$ of subsets of some set X each satisfy the conditions of *Theorem* (1.2.5) and hence generate unique topologies τ_1 and τ_2. If $\tau_1 = \tau_2$ then $\mathcal{B}_1$ and $\mathcal{B}_2$ are called equivalent bases. The question to be answered is "what conditions on $\mathcal{B}_1$ and $\mathcal{B}_2$ will guarantee that $\mathcal{B}_1$ and $\mathcal{B}_2$ are equivalent bases?" The question is answered by the following theorem. Once again we refer the reader to Greever [2].

Theorem 1.2.7*

Let X be a set and $\mathcal{B}_1$ and $\mathcal{B}_2$ bases for topologies τ_1 and τ_2 on X. If

i) $x \in B_1 \in \mathcal{B}_1$ implies that there exists $B_2 \in \mathcal{B}_2$ such that $x \in B_2 \subseteq B_1$ and

ii) $x \in B_2 \in \mathcal{B}_2$ implies that there exists $B_1 \in \mathcal{B}_1$ such that $x \in B_1 \subseteq B_2$ then $\mathcal{B}_1$ and $\mathcal{B}_2$ are equivalent bases.

Theorem 1.2.8

$\mathcal{B}_1 = \{(d_1, d_2) : d_1$ and d_2 are rational$\}$ is equivalent to the base $\mathcal{B}$ of *Definition* (1.2.6) for the Euclidean topology $\mathcal{E}$ on $\mathcal{R}$.

There are situations in which we have at hand a topological space (X, τ) and a subset A of X. It is possible to use the topology τ to induce a topology τ_A on A by the following technique.

Definition 1.2.9

Let (X, τ) be a topological space and let $A \subseteq X$. Let

$$\tau_A = \{A \cap U : U \in \tau\}.$$

τ_A is called the *relative topology* for A.

It is of interest to observe that if (X, τ) is a topological space with base $\mathcal{B}_1$ and if $A \subseteq X$, then the collection $\mathcal{B}_2 = \{B \cap A : B \in \mathcal{B}_1\}$ serves as a base for the topology τ_A of *Definition* (1.2.9).

Definition 1.2.10

A topological space (X, τ) is *second countable* if and only if there exists a countable base for τ.

Theorem 1.2.11

$(\mathcal{R}, \mathcal{E})$ is second countable.

Definition 1.2.12

Let X and Y be sets and let f be a function mapping X into Y, denoted $f : X \to Y$. Let $A \subseteq Y$. The *inverse image* of A under f denoted $f^{-1}(A)$ is defined by

$$f^{-1}(A) = \{x \in X : f(x) \in A\}.$$

The following theorem summarizes some of the properties of inverse images and will be of use in connection with our study of measurable functions and random variables.

Theorem 1.2.13

Let $f : X \to Y$. Then

i) $f^{-1}(\phi) = \phi$;

ii) $f^{-1}(Y) = X$;

iii) if $A, B, \subseteq Y$, $f^{-1}(A - B) = f^{-1}(A) - f^{-1}(B)$;

iv) if $\{A_\gamma : \gamma \in \Gamma\}$ is a collection of subsets of Y then $f^{-1}(\bigcup_{\gamma\in\Gamma} A_\gamma) = \bigcup_{\gamma\in\Gamma} f^{-1}(A_\gamma)$

$$f^{-1}(\bigcap_{\gamma\in\Gamma} A_\gamma) = \bigcap_{\gamma\in\Gamma} f^{-1}(A_\gamma).$$

Proof: i) and ii) are obvious from the definition of f^{-1}. To prove iii), let $x \in f^{-1}(A - B)$. Then $f(x) \in A - B$ implying that $f(x) \in A$ but $f(x) \notin B$. Hence $x \in f^{-1}(A)$ but $x \notin f^{-1}(B)$. This simply says that $x \in f^{-1}(A) - f^{-1}(B)$. Hence $f^{-1}(A - B) \subseteq f^{-1}(A) - f^{-1}(B)$. Now let $x \in f^{-1}(A) - f^{-1}(B)$. Then $f(x) \in A$ but $f(x) \notin B$. Thus $f(x) \in A - B$ implying that $x \in f^{-1}(A - B)$. Hence $f^{-1}(A) - f^{-1}(B) \subseteq f^{-1}(A - B)$. By *Definition* (1.1.6), $f^{-1}(A - B) = f^{-1}(A) - f^{-1}(B)$. To verify iv), let $x \in f^{-1}(\bigcup_{\gamma\in\Gamma} A_\gamma)$. Then $f(x) \in \bigcup_{\gamma\in\Gamma} A_\gamma$ implying that $f(x) \in A_\gamma$ for at least one $\gamma \in \Gamma$. Hence $x \in f^{-1}(A_\gamma)$ for at least one $\gamma \in \Gamma$. Thus by *Definition* (1.1.11) $x \in \bigcup_{\gamma\in\Gamma} f^{-1}(A_\gamma)$ and we have that $f^{-1}(\bigcup_{\gamma\in\Gamma} A_\gamma) \subseteq \bigcup_{\gamma\in\Gamma} f^{-1}(A_\gamma)$. Now let $x \in \bigcup_{\gamma\in\Gamma} f^{-1}(A_\gamma)$. Then $x \in f^{-1}(A_\gamma)$ for at least one $\gamma \in \Gamma$ implying that $f(x) \in A_\gamma$ for at least one $\gamma \in \Gamma$. Hence $f(x) \in \bigcup_{\gamma\in\Gamma} A_\gamma$ by *Definition* (1.1.11). This in turn implies that $x \in f^{-1}(\bigcup_{\gamma\in\Gamma} A_\gamma)$ and that $\bigcup_{\gamma\in\Gamma} f^{-1}(A_\gamma) \subseteq f^{-1}(\bigcup_{\gamma\in\Gamma} A_\gamma)$. By *Definition* (1.1.6) $f^{-1}(\bigcup_{\gamma\in\Gamma} A_\gamma) = \bigcup_{\gamma\in\Gamma} f^{-1}(A_\gamma)$. To see that iv) is true for set intersection let $x \in f^{-1}(\bigcap_{\gamma\in\Gamma} A_\gamma)$. Then $f(x) \in \bigcap_{\gamma\in\Gamma} A_\gamma$ implying that $f(x) \in A_\gamma$ for each $\gamma \in \Gamma$ by *Definition*

(1.1.12). This in turn implies that $x \in f^{-1}(A_\gamma)$ for each $\gamma \in \Gamma$ or that $x \in \bigcap_{\gamma\in\Gamma} f^{-1}(A_\gamma)$. Hence we obtain that $f^{-1}(\bigcap_{\gamma\in\Gamma} A_\gamma) \subseteq \bigcap_{\gamma\in\Gamma} f^{-1}(A_\gamma)$. Now let $x \in \bigcap_{\gamma\in\Gamma} f^{-1}(A_\gamma)$. Then $x \in f^{-1}(A_\gamma)$ for each $\gamma \in \Gamma$ implying that $f(x) \in A_\gamma$ for each $\gamma \in \Gamma$ and that $f(x) \in \bigcap_{\gamma\in\Gamma} A_\gamma$. This in turn implies that $x \in f^{-1}(\bigcap_{\gamma\in\Gamma} A_\gamma)$ and that $\bigcap_{\gamma\in\Gamma} f^{-1}(A_\gamma) \subseteq f^{-1}(\bigcap_{\gamma\in\Gamma} A_\gamma)$.

By *Definition* (1.1.6) $f^{-1}(\bigcap_{\gamma\in\Gamma} A_\gamma) = \bigcap_{\gamma\in\Gamma} f^{-1}(A_\gamma)$.

Definition 1.2.14

Let (X, τ) and (Y, τ_1) be topological spaces and let $f : X \to Y$. f is said to be *continuous at a point* $x_0 \in X$ if, given any set $V \in \tau_1$ such that $f(x_0) \in V$ there exists a set $U \in \tau$ such that $x_0 \in U$ and for each $x \in U$, $f(x) \in V$.

Definition 1.2.15

We say that f is *continuous* if and only if f is continuous at each point of X and refer to this as *pointwise continuity.*

The following theorem presents a useful characterization of continuous functions from one topological space to another and can be used as an equivalent form for *Definition* (1.2.15).

Theorem 1.2.16

Let (X, τ) and (Y, τ_1) be topological spaces and let $f : X \to Y$. f is continuous if and only if $f^{-1}(V) \in \tau$ for each $V \in \tau_1$.

Proof: Assume that f is continuous from X into Y. Let $V \in \tau_1$. Let $A = \{a_\gamma \in V : a_\gamma = f(x_\gamma)$ for some $x_\gamma \in X, \gamma \in \Gamma\}$. By *Definition* (1.2.14) for each $a_\gamma \in A$ there exists an open set U_γ such that $x_\gamma \in U_\gamma$ and such that $x \in U_\gamma$ implies $f(x) \in V$. Consider $\bigcup_{\gamma \in \Gamma} U_\gamma$ and note that by *Definition* (1.2.1) this set is an element of τ. We claim that $\bigcup_{\gamma \in \Gamma} U_\gamma = f^{-1}(V)$. To verify this, let $x \in \bigcup_{\gamma \in \Gamma} U_\gamma$. Then $x \in U_\gamma$ for some $\gamma \in \Gamma$. By choice of U_γ, $f(x) \in V$ implying that $x \in f^{-1}(V)$. Hence $\bigcup_{\gamma \in \Gamma} U_\gamma \subseteq f^{-1}(V)$. Now let $x \in f^{-1}(V)$. Then $f(x) \in V$ implying that $f(x) = a_\gamma$ for some $a_\gamma \in A$. Thus we can let $x = x_\gamma$ which implies that $x \in U_\gamma$ for some $\gamma \in \Gamma$ and that $x \in \bigcup_{\gamma \in \Gamma} U_\gamma$. Hence $f^{-1}(V) \subseteq \bigcup_{\gamma \in \Gamma} U_\gamma$ and we have by *Definition* (1.1.6) that $\bigcup_{\gamma \in \Gamma} U_\gamma = f^{-1}(V)$. Now assume that the inverse image of any set $V \in \tau_1$ is an element of τ. Let $x_0 \in X$. Let $V \in \tau_1$ be such that $f(x_0) \in V$. Consider $f^{-1}(V)$. $x_0 \in f^{-1}(V)$ and by assumption $f^{-1}(V) \in \tau$. $f^{-1}(V)$ also satisfies the condition that for each $x \in f^{-1}(V)$, $f(x) \in V$ as is required in *Definition* (1.2.14).

Note 1.2.17

We shall be particularly concerned with real valued functions f whose domain is some set $A \subseteq R$. In this case when we write "f is continuous" we mean that $f : A \to R$ and f is continuous at each point of A with respect to the topological spaces (A, τ_A) of *Definition* (1.2.9) and (R, E) of *Definition* (1.2.6).

Definition 1.2.18

Let (X, τ) be a topological space, (x_n) a sequence in X and $x_0 \in X$. The sequence (x_n) converges to x_0 denoted $(x_n) \to x_0$ if given any set $U \in \tau$ such that $x_0 \in U$ there exists a natural number N_1 such that $n > N_1$ implies $x_n \in U$.

Note 1.2.19

In the case of the topological space (R, E), *Definition* (1.2.18) takes the following form which is the usual form seen in elementary calculus courses:

"Let (x_n) be a sequence of real numbers and let $x_0 \in R$. Then $(x_n) \to x_0$ if and only if given any $\varepsilon > 0$ there exists a natural number N_1 such that for $n > N_1$

$$|x_n - x_0| < \varepsilon."$$

The following theorem presents an equivalent and useful form for *Definition* (1.2.14) in the special case in which f is a real valued function whose domain is some set $A \subseteq R$.

Theorem 1.2.20

Let f be a real valued function with domain $A \subseteq R$. f is continuous at $x_0 \in A$ if and only if given any sequence of points (x_n) in A such that $(x_n) \to x_0$, $(f(x_n)) \to f(x_0)$.

Proof: Assume that f is continuous at x_0 and that

$(x_n) \to x_0$. Let $\varepsilon > 0$ and let $V = (f(x_0) - \varepsilon, f(x_0) + \varepsilon)$. Since f is continuous at x_0 and since $V \in \mathcal{E}$, by *Definition* (1.2.14) and *Note* (1.2.17), there exists a set $U \in \mathcal{E}$ such that $x_0 \in U \cap A$ and $f(x) \in V$ for each $x \in U \cap A$. Since $(x_n) \to x_0$, by *Definition* (1.2.18) there exists a natural number N_1 such that $n > N_1$ implies $x_n \in U \cap A$. Hence for $n > N_1$, $f(x_n) \in V$. This implies that for $n > N_1$, $|f(x_n) - f(x_0)| < \varepsilon$ which by *Note* (1.2.19) is sufficient to show that

$$(f(x_n)) \to f(x_0).$$

To prove the converse, assume that given any sequence (x_n) of points in A which converge to x_0, $(f(x_n)) \to f(x_0)$ but that f is not continuous at x_0. Since f is not continuous at x_0 there exists a basic open set $V = (f(x_0) - \varepsilon, f(x_0) + \varepsilon)$ such that for each element $A \cap U \in \tau_A$ containing x_0 where $U \in \mathcal{E}$ there exists an element $x \in A \cap U$ such that $f(x) \notin V$. In particular consider the sequence of open sets

$$(A \cap (x_0 - 1/m, x_0 + 1/m)), \ m = 1, 2, 3, 4, \ldots$$

Let $x_m' \in A \cap (x_0 - 1/m, x_0 + 1/m)$ be such that $f(x_m') \notin V$. We claim that $(x_m') \to x_0$. To verify this let Z be any open set about x_0. There exists a basic open set of the form

$$A \cap (x_0 - c, x_0 + c) \quad \text{for some} \quad c > 0$$

such that

$$A \cap (x_0 - c, x_0 + c) \subseteq Z.$$

Since

$$\lim_{m \to \infty} 1/m = 0,$$

there exists a natural number M such that $m > M$ implies $1/m < c$. Hence also for $m > M$, $(x_0 - 1/m, x_0 + 1/m) \subseteq (x_0 - c, x_0 + c)$ and $A \cap (x_0 - 1/m, x_0 + 1/m \subseteq A \cap (x_0 - c, x_0 + c) \subseteq Z$. This in turn implies that for $m > M$, $x_m' \in Z$ as was desired. By assumption, $(x_m') \to x_0$ implies that $f(x_m') \to f(x_0)$. Hence there exists a natural number M_1 such that $m > M_1$ implies

$$|f(x_0) - f(x_m')| < \varepsilon$$

by *Note* (1.2.19). Hence for $m > M_1$,

$$f(x_m') \in V$$

which is a contradiction and the proof is complete.

Definition 1.2.21

Let f be a real valued function with domain $A \subseteq R$. Let $x_0 \in R$. The *right hand limit* of f at x_0 is b, denoted $\lim_{x \to x_0^+} f(x) = b$ if and only if given any $\varepsilon > 0$ there exists a $\delta > 0$ such that $(x_0, x_0 + \delta) \subseteq A$ and such that $x \in (x_0, x_0 + \delta)$ implies $f(x) \in (b - \varepsilon, b + \varepsilon)$.

When we write $\lim_{x \to x_0^+} f(x) = b$ we are assuming of course that $b \in R$ and hence that this limit exists and is finite. A similar definition can be stated for the left hand limit of f at x_0.

It is now possible to define the concept of right hand continuity for real valued functions f whose domain is some set $A \subseteq R$. This idea will play an important role in our discussion of distribution functions in Chapter Four.

Definition 1.2.22

Let f be a real valued function with domain $A \subseteq R$. Let $x_0 \in R$. f is said to be *continuous from the right* at x_0 if and only if

i) $x_0 \in A$;

and

ii) $\lim_{x \to x_0^+} f(x) = f(x_0)$.

As was the case when dealing with continuity of f at x_0 it is useful to find an equivalent form for *Definition* (1.2.22) in sequential terms. The following theorem provides one such equivalency.

Theorem 1.2.23

Let f be a real valued function with domain $A \subseteq R$. Let $[s, t) \subseteq A$ and let $x_0 \in [s, t)$. The following are equivalent:

i) f is continuous from the right at x_0;

ii) given any sequence (x_n) of elements of A such that $x_n > x_0$ for each n and $(x_n) \to x_0$, $(f(x_n)) \to f(x_0)$;

iii) given any sequence (x_n) of elements of A such that $x_1 > x_2 > x_3 > \cdots > x_n > \cdots > x_0$ and $(x_n) \to x_0$, $(f(x_n)) \to f(x_0)$.

Proof: We shall prove the above by showing that i) → ii) → iii) → i). i) → ii). Assume that f is continuous from the right at x_0. Let (x_n) be a sequence of elements of A satisfying the condition of statement ii). Choose $\varepsilon > 0$ and consider $V = (f(x_0) - \varepsilon, f(x_0) + \varepsilon)$. By *Definition* (1.2.22), $\lim_{x \to x_0^+} f(x) = f(x_0)$ and hence by *Definition* (1.2.21) there exists a $\delta > 0$ such that $(x_0, x_0 + \delta) \subseteq A$ and such that for $x \in (x_0, x_0 + \delta)$ $f(x) \in V$. Consider $(x_0 - \delta, x_0 + \delta) \cap A \in \tau_A$. Since $(x_n) \to x_0$ there exists a natural number N_1 such that $n > N_1$ implies $x_n \in (x_0 - \delta, x_0 + \delta) \cap A$. Since by assumption $x_0 < x_n$, we have that for $n > N_1$, $x_n \in (x_0, x_0 + \delta) \cap A$ which in turn implies that for $n > N_1$,

$f(x_n) \in V$. Thus for $n > N_1$, $|f(x_n) - f(x_0)| < \varepsilon$ and by *Note* (1.2.19) we have that $(f(x_n)) \to f(x_0)$. The fact that ii) $\to$ iii) is obvious. To show that iii) $\to$ i) we shall use the method of contradiction. Assume that f is not continuous from the right at x_0 but that given any sequence (x_n) of elements of A such that $(x_n) \to x_0$ and $x_n > x_0$ for each n, $(f(x_n)) \to f(x_0)$. There exists an $\varepsilon > 0$ such that for each interval of the form $(x_0, x_0 + \delta) \subseteq A$ there exists an x such that $x \in (x_0, x_0 + \delta)$ but $f(x) \notin V$ where $V = (f(x_0) - \varepsilon, f(x_0) + \varepsilon)$. Consider first the sequence of sets $(x_0, x_0 + 1/q)$, $q = 1, 2, 3, \ldots$. There exists a smallest natural number Q such that $q \geq Q$ implies $(x_0, x_0 + 1/q) \subseteq A$. Now consider the sequence of sets $(x_0, x_0 + 1/m)$, $m = Q, Q + 1, Q + 2, \ldots$. For each m there exists an element $x_m' \in (x_0, x_0 + 1/m)$ such that $f(x_m') \notin V$. We claim that $(x_m') \to x_0$. To see this let $x_0 \in A \cap U$ for $U \in \mathcal{E}$. There exists a $c > 0$ such that $(x_0, x_0 + c) \subseteq A \cap U$. Since $\lim_{m\to\infty} 1/m = 0$, there exists a natural number M such that $m > M$ implies $1/m < c$. Thus for $m > M$, $(x_0, x_0 + 1/m) \subseteq (x_0, x_0 + c) \subseteq A \cap U$. This in turn implies that for $m > M$, $x_m' \in A \cap U$ as was desired. By assumption $(f(x_m')) \to f(x_0)$. Thus there exists a natural number M_1 such that $m > M_1$ implies $|f(x_m') - f(x_0)| < \varepsilon$ by *Note* (1.2.19). Hence for $m > M_1$, $f(x_m') \in V$ which is a contradiction and the proof is complete.

Thus far we have been concerned principally with the usual topology of the real line and with properties of maps from $\mathcal{R}$ into $\mathcal{R}$. It will be necessary to consider briefly what is known as Euclidean n-space for $n \in N$ as this concept will be basic in Chapter Six when we consider what will be termed n-dimensional random variables. Before we define n-space it will be necessary to define the term "Cartesian product" from set theory. This concept has been mentioned previously in *Theorem* (1.1.13) but was not defined formally at that time.

Definition 1.2.24

Let $A_1, A_2, \ldots, A_n$ be non-empty sets. The *Cartesian product* of $A_1, A_2, \ldots, A_n$ denoted $A_1 \times A_2 \times A_3 \times \ldots \times A_n = \overset{n}{\underset{i=1}{\times}} A_i$ is the set $\{(a_1, a_2, \ldots, a_n): a_1 \in A_1, a_2 \in A_2, \ldots, a_n \in A_n\}$. If $A_1 = A_2 = \cdots = A_n$ we shall denote the Cartesian product simply by A^n.

Theorem 1.2.25

Let $\mathcal{R}_+$ denote the positive real numbers. Let $N_\varepsilon(p)$ be as in *Definition* (1.2.6). Let

$$\mathcal{B} = \{B: B = N_{\varepsilon_1}(p_1) \times N_{\varepsilon_2}(p_2) \times \cdots N_{\varepsilon_n}(p_n) \text{ for some } (p_1, p_2, \ldots, p_n) \in \mathcal{R}^n \text{ and } (\varepsilon_1, \varepsilon_2, \ldots, \varepsilon_n) \in \mathcal{R}_+^n\}.$$

$\mathcal{B}$ is a base for a unique topology τ on $\mathcal{R}^n$.

Proof: We shall show that conditions i) and ii) of *Theorem* (1.2.5) are satisfied. Let $x \in \mathcal{R}^n$. By definition $x = (x_1, x_2, \ldots, x_n)$ where for each i x_i is real. Choose $\varepsilon > 0$ and consider

$$B = N_\varepsilon(x_1) \times N_\varepsilon(x_2) \times \cdots N_\varepsilon(x_n).$$

$B \in \mathcal{B}$ and $x \in B$. Thus i) of *Theorem* (1.2.5) is easily satisfied. Now let B_1 and $B_2 \in \mathcal{B}$ and let $x \in B_1 \cap B_2$. Let $x = (x_1, x_2, x_3, \ldots, x_n)$. By definition

$B_1 = \overset{n}{\underset{i=1}{\times}} N_{\varepsilon_i}(p_i)$ for p_i real and $\varepsilon_i > 0$ and

$B_2 = \overset{n}{\underset{i=1}{\times}} N_{\varepsilon_i'}(p_i')$ for p_i' real and $\varepsilon_i' > 0$.

$x \in B_1 \cap B_2$ implies that for each i

$x_i \in N_{\varepsilon_i}(p_i) \cap N_{\varepsilon_i'}(p_i')$. Since the intersection of two open intervals about a given point is also an open interval about that point we have that $N_{\varepsilon_i}(p_i) \cap N_{\varepsilon_i'}(p_i')$ is an open interval about x_i and also an open set about x_i with respect to the usual topology on $\mathcal{R}$. Thus $N_{\varepsilon_i}(p_i) \cap N_{\varepsilon_i'}(p_i')$ contains a basic open set $N_{\sigma_i}(x_i)$ about x_i. Let $B = \overset{n}{\underset{i=1}{\times}} N_{\sigma_i}(x_i)$. It is obvious that $B \in \mathcal{B}$ and $B \subseteq B_1 \cap B_2$. Hence by *Theorem* (1.2.5) $\mathcal{B}$ is the base for a unique topology τ on $\mathcal{R}^n$.

Definition 1.2.26

The topology τ on R^n for which $\mathcal{B}$ of *Theorem* (1.2.25) is the base is called the Euclidean topology on R^n. We shall denote this topology by E^n. When we speak of Euclidean n-space we shall mean the topological space (R^n, E^n).

Note 1.2.27

If $n = 1$ then obviously $(R^n, E^n) = (R, E)$ which has been discussed previously. Note also that *Definition* (1.2.14), *Definition* (1.2.15) and *Theorem* (1.2.16) hold for general topological spaces (X, τ) and will hold in particular for the space (R^n, E^n).

1.2 *EXERCISES*

1. Give an example of an ordered pair (X, τ) which is a topological space but for which τ is <u>not</u> a σ algebra.
2. Give an example of a non-empty set X and a collection C of subsets of X such that C is a σ algebra but (X, C) is not a topological space.
3. Give an example of a non-empty set X and a collection τ of subsets of X such that (X, τ) is a topological space and τ is also a σ algebra.
4. Let $f: X \to Y$. Let $A \subseteq X$ and let $B \subseteq Y$. Prove each of the following:

   i) $A \subseteq f^{-1}(f(A))$ where $f(A) = \{y: y = f(x)$ for some $x \in A\}$;

   ii) In general $A \neq f^{-1}(f(A))$;

   iii) $f(f^{-1}(B)) \subseteq B$;

   iv) If B is a subset of the range of f then

$f(f^{-1}(B)) = B.$

5. Let $X = R$. Let H be the collection of all non-empty subsets U of R such that for each point $p \in U$ there exists an interval of the form $[a, b)$, $a < b$ such that $p \in [a, b)$ and $[a, b) \subseteq U$. Let $\tau_1 = \{\phi\} \cup H$. Verify that (X, τ_1) is a topological space.

6. Let $X = R$. Let τ_1 be as defined in *Exercise* (1.2.5). Let $\tau_2 = E$ (see *Definition* (1.2.6)). Let $\tau_3 = \rho_X$ (see *Definition* (1.1.33)). Let $\tau_4 = \{\phi, X\}$. Each of the ordered pairs (X, τ_i) $i = 1, 2, 3, 4$ is a topological space. Consider the functions f and g defined on R as follows:

$$f(x) = \begin{cases} -1 & \text{if } x < 0 \\ 1 & \text{if } x \geq 0 \end{cases}$$

$$g(x) = \begin{cases} 0 & \text{if } x \text{ is rational} \\ 1 & \text{if } x \text{ is irrational} \end{cases}$$

Let $(X, \tau_i) \rightarrow (X, \tau_j)$ indicate that the given function is continuous from (X, τ_i) into (X, τ_j) and let $(X, \tau_i) \not\rightarrow (X, \tau_j)$ indicate that continuity fails at some point. Verify the relationship indicated in the following diagrams.

Continuity Diagram for the function f

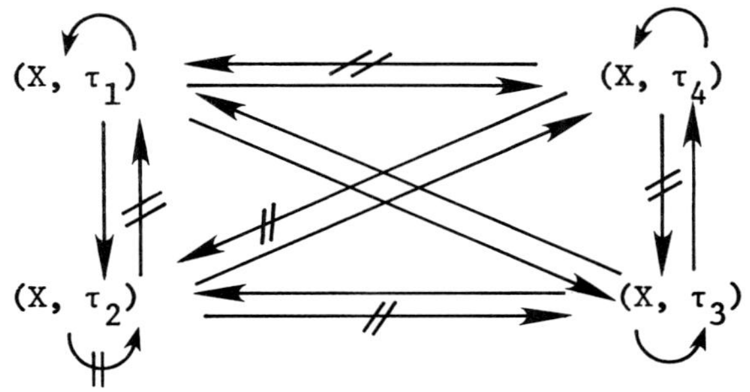

Continuity Diagram for the function g

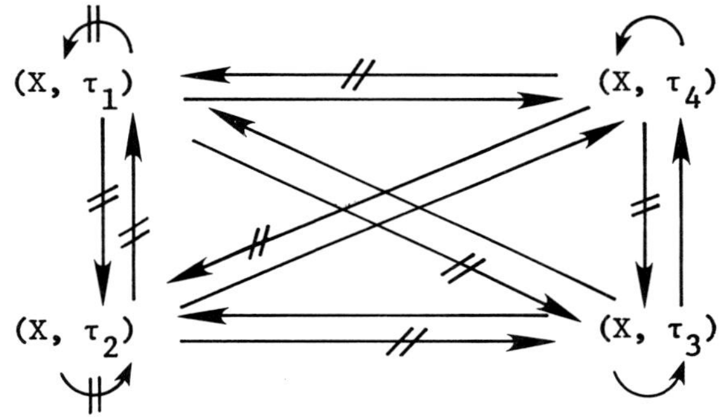

1.3 *SUMMARY*

The basic concepts of *elementary set theory* were developed and the properties of the usual set theoretic operations of *union*, *intersection*, *complementation*, and *limits* were investigated. Two types of mathematical systems which concern collections of sets and the above set theoretic operations were presented. These were the following:

Let X be a set and let

$$\mathcal{C} = \{A_\gamma : \gamma \in \Gamma\}$$

be a collection of subsets of X. $\mathcal{C}$ is said to be a *σ-algebra* if the following properties hold:

i) $\phi \in \mathcal{C}$;

ii) If $A \in \mathcal{C}$, then $X - A \in \mathcal{C}$;

iii) The union of any countable collection of elements of $\mathcal{C}$ is an element of $\mathcal{C}$.

A *topological space* is an ordered pair (X, τ) where X is a set and τ is a collection of subsets of X such that

i) $\phi \in \tau$ and $X \in \tau$;

ii) $U \cap V \in \tau$ whenever U and $V \in \tau$;

iii) $\bigcup_{\gamma \in \Gamma} U_\gamma \in \tau$ whenever $\{U_\gamma : \gamma \in \Gamma\} \subseteq \tau$.

τ is called a *topology* for X and the elements of τ are called *open sets*.

The *Euclidean topology* for the real line was studied in some detail and particular emphasis was placed on questions of continuity.

IMPORTANT TERMS IN CHAPTER ONE

subset
universal set
null set
countable set
set union
set intersection
set complement
disjoint sets
σ-algebra
lim sup A_n
lim inf A_n
power set
partition
topological space
base for a topology
Euclidean topology
relative topology
second countable
inverse image of A *under* f
pointwise continuity
convergence of a sequence
right hand limit
right hand continuity
Cartesian product

REFERENCES AND SUGGESTIONS FOR FURTHER READINGS

[1] Bartle, R. G., The Elements of Real Analysis. New York: John Wiley and Sons, Inc., 1964.

[2] Greever, John, Theory and Examples of Point Set Topology. Belmont, California: Wadsworth Publishing Company, Inc., 1967.

[3] **Halmos, Paul R., Naive Set Theory, Princeton, New Jersey: D. Van Nostrand Company, Inc., 1960.

[4] Kelley, John L., General Topology. Princeton, New Jersey: D. Van Nostrand Company, Inc., 1955.

[5] Royden, H. L., Real Analysis. New York: The MacMillan Company, 1963.

[6] Rudin, Walter, Principles of Mathematical Analysis. New York: McGraw Hill Book Company, 1964.

[7] **Taylor, Angus E., General Theory of Functions and Integration. New York: Blaisdell Publishing Company, 1965.

**These books are more advanced than the approach of the present text.

CHAPTER TWO

ELEMENTARY MEASURE THEORY, LEBESGUE AND RIEMANN-STIELTJES INTEGRAL

2.0 *INTRODUCTION*

As we mentioned in Chapter One our aim in this chapter is to review some basic mathematical concepts which will be needed in our approach to the study of *probability theory*. We shall discuss some *elementary aspects of measure theory, the Lebesgue integral, the Riemann-Stieltjes integral* and *some miscellaneous topics from analysis*.

2.1 *ELEMENTARY MEASURE THEORY*

We shall present in this section some of the basic concepts of measure theory. These results will provide the foundation for our later study of probability spaces and random variables.

Definition 2.1.1

An ordered pair $(X, \mathcal{C})$ such that X is a set and $\mathcal{C}$ a σ algebra of subsets of X is called a *measurable space*. Any set $A \in \mathcal{C}$ is said to be measurable.

Example 2.1.2

In each case of *Example* (1.1.24) the ordered pair $(X, \mathcal{C})$ is a measurable space.

Example 2.1.3

Let $(X, \mathcal{C})$ be defined such that X = real numbers and $\mathcal{C}$ is the σ algebra generated by all open intervals (a, b) of the real line. $\mathcal{C}$ is called the Borel algebra and any set in $\mathcal{C}$ is called a Borel set. We shall denote the Borel sets by β. The Borel sets are often defined to be the σ algebra generated by the open sets of $\mathcal{R}$ relative to the usual topology. It can be shown that these definitions are equivalent.

Note 2.1.4

By the extended reals we mean Reals $\cup\{-\infty, \infty\} = \overline{\mathcal{R}}$. The symbols $-\infty$ and ∞ are not to be considered as real numbers. This is simply a notational convenience which will allow us to write such expressions as $-\infty < x < \infty$ for x real and which will render meaningful statements such as "the length of the real line is equal to ∞." We shall also be employing the convention that $0(+\infty) = 0$. This will prove useful in the development of the Lebesgue integral in section two.

Definition 2.1.5

A function mapping into the extended reals whose domain is a collection of sets is called a *set function*.

Example 2.1.6

i) Let X be the set N of natural numbers and let $\mathcal{C}$ be the collection of all non-empty subsets of N. Define a map $f : \mathcal{C} \to \mathcal{R}$ by f(A) = least element of A for $A \in \mathcal{C}$. f is a set function.

ii) Let X be the set of real numbers and let $\mathcal{C}$ be the collection of all open intervals (a, b) of $\mathcal{R}$. Define a function $f : \mathcal{C} \to \mathcal{R}$ by $f((a, b)) = b - a$. f is a set function.

iii) Let X be an uncountable set and $\mathcal{C}$ the collection of all subsets of X. Define a function f on $\mathcal{C}$ by $f(A) = 0$ if A is countable and $f(A) = \infty$ if A is uncountable. f is a set function.

iv) Let X be the set N of natural numbers. Let $\mathcal{C}$ be the collection of all subsets of N. Define a function f on $\mathcal{C}$ by $f(A) = 0$ if A is finite and $f(A) = \infty$ if A is infinite. f is a set function.

Definition 2.1.7

Let $(X, \mathcal{C})$ be a measurable space. A *measure* is a set function λ defined on $\mathcal{C}$ such that

i) $\lambda(\phi) = 0$;

ii) $\lambda(A) \geq 0$ for each $A \in \mathcal{C}$;

iii) λ is countably additive in the sense that if (A_n) is a sequence of elements of $\mathcal{C}$ such that $A_i \cap A_j = \phi$ for $i \neq j$ then

$$\lambda\left(\bigcup_{n=1}^{\infty} A_n\right) = \sum_{n=1}^{\infty} \lambda(A_n).$$

Definition 2.1.8

An ordered triple $(X, \mathcal{C}, \lambda)$ where $(X, \mathcal{C})$ is a measurable space and λ is a measure on $\mathcal{C}$ is called a *measure space*.

Example 2.1.9

Referring to the examples given in *Example* (2.1.6), $(X, \mathcal{C}, f)$ as defined in iii) is a measure space but $(X, \mathcal{C}, f)$ as defined in iv) is not.

Proof: To check iii) is straightforward. To see that iv) fails consider

$$A_1 = \{1\}$$
$$A_2 = \{2\}$$
$$A_3 = \{3\}$$
$$\vdots$$
$$A_n = \{n\}$$

Then $\bigcup_{n=1}^{\infty} A_n = N$ which is an infinite set. Thus

$$f\left(\bigcup_{n=1}^{\infty} A_n\right) = \infty \neq \sum_{n=1}^{\infty} f(A_n) = 0.$$

Definition 2.1.10

Let $(X, \mathcal{C}, \lambda)$ be a measure space. If for each $A \in \mathcal{C}$, $\lambda(A) < \infty$ then λ is said to be *finite* and we refer to $(X, \mathcal{C}, \lambda)$ as a *finite measure space*.

Definition 2.1.11

Let $(X, \mathcal{C}, \lambda)$ be a measure space. If there exists a sequence (A_n) of elements of $\mathcal{C}$ such that

$$\lambda(A_i) < \infty \quad \text{for each} \quad i = 1, 2, 3, \ldots \quad \text{and}$$

$$\bigcup_{i=1}^{\infty} A_i = X$$

then λ is said to be σ finite and we refer to $(X, \mathcal{C}, \lambda)$ as a σ *finite measure space.*

Example 2.1.12

Let X be the set N of natural numbers and $\mathcal{C}$ be the collection of all subsets of N. Define a map f on $\mathcal{C}$ by $f(A)$ = the number of elements in A for A finite; $f(A) = \infty$ for A infinite. Then $(X, \mathcal{C}, f)$ is a measure space which is not finite but it is σ finite.

We wish to consider briefly the topic of Lebesgue measure on the real line. This concept will provide an insight into the general structure of a measure space as well as provide some useful examples for later chapters.

The aim of Lebesgue measure is to define a measure space $(\mathcal{R}, F^*, \ell^*)$ with the properties that F^* will contain every finite interval I and such that $\ell^*(I)$ = length I. We shall present here a description of the process used to define Lebesgue measure. No attempt will be made to prove the lemmas or theorems stated here. For detailed proofs the reader is referred to Bartle [2].

Definition 2.1.13

By the length of intervals of the forms (a, b), $(a, b]$, $[a, b)$, $a \leq b$, we shall mean the *real number* $b - a$.

By the length of the infinite intervals of the form $(-\infty, b]$, (a, ∞), $(-\infty, \infty)$ we shall mean the *extended real number* ∞. We shall define the length of the union of a finite number of disjoint sets of these forms to be the sum of the corresponding lengths.

Definition 2.1.14

Let X be a set and let $\mathcal{C} = \{A_\gamma : \gamma \in \Gamma\}$ be a collection of subsets of X. $\mathcal{C}$ is said to be an *algebra* if

i) $\phi \in \mathcal{C}$;

ii) If $A \in \mathcal{C}$, then $X - A \in \mathcal{C}$

iii) If $A_1, A_2, \ldots A_n \in \mathcal{C}$ then $\bigcup_{i=1}^{n} A_i \in \mathcal{C}$.

**Lemma* 2.1.15

The collection F of all finite unions of sets of the form $(a, b]$, $(-\infty, b]$, (a, ∞), $(-\infty, \infty)$ where $a \leq b$ is an algebra of sets of $\mathcal{R}$.

Note 2.1.16

It is evident from *Definition* (2.1.14) and *Definition* (1.1.22) that any σ algebra of sets of X is also an algebra of sets but the converse need not be true as can be seen from the following example.

Example 2.1.17

The collection F of *Lemma* (2.1.15) does not form a σ algebra of sets. This can be seen by considering the fact that $(a, b) \notin F$ but

$$(a, b) = \bigcup_{n=1}^{\infty} (a, b - \frac{1}{n}].$$

Definition 2.1.18

Let X be a set and $\mathcal{C}$ an algebra of subsets of X. A measure on $\mathcal{C}$ is a *set function* λ defined on $\mathcal{C}$ such that

i) $\lambda(\phi) = 0$;

ii) $\lambda(A) \geq 0$ for all $A \in \mathcal{C}$;

iii) λ is countably additive in the sense that if (A_n) is a sequence of elements of $\mathcal{C}$ such that $A_i \cap A_j = \phi$, $i \neq j$ and $\bigcup_{n=1}^{\infty} A_n \in \mathcal{C}$ then

$$\lambda\left(\bigcup_{n=1}^{\infty} A_n\right) = \sum_{n=1}^{\infty} \lambda(A_n).$$

**Lemma* 2.1.19

Length is a measure on the algebra F of *Lemma* (2.1.15). We shall denote the length measure on F by ℓ.

The extension technique to be employed consists of showing that in general it is possible to take any algebra $\mathcal{C}$ of subsets of some set X and any measure λ defined on $\mathcal{C}$ and construct a measure space $(X, \mathcal{C}^*, \lambda^*)$ which will have the properties that $\mathcal{C} \subseteq \mathcal{C}^*$ and $\lambda^*(A) = \lambda(A)$ for any $A \in \mathcal{C}$. This process applied to the algebra F with measure ℓ defined on F will yield a measure space $(\mathcal{R}, F^*, \ell^*)$. F^* is commonly referred to as the collection of Lebesgue measurable sets and ℓ^* is called Lebesgue measure. We outline below the extension procedure.

Definition 2.1.20

Let X be a set, $\mathcal{C}$ be an algebra of subsets of X, and λ a measure on $\mathcal{C}$. Let $B \subseteq X$. We define the *outer measure* of B generated by λ denoted $\lambda^*(B)$ by

$$\lambda^*(B) = \inf \sum_{n=1}^{\infty} \lambda(A_n)$$

where the infimum is extended over all sequences (A_n) of elements of $\mathcal{C}$ such that $B \subseteq \bigcup_{n=1}^{\infty} A_n$.

Note 2.1.21

Since $X \in \mathcal{C}$ and $B \subseteq X$, $\{\sum_{n=1}^{\infty} \lambda(A_n) : B \subseteq \bigcup_{n=1}^{\infty} A_n\} \neq \phi$. Note also that by definition of a measure on an algebra, this set is bounded below by 0. Hence by the completeness property of the real numbers, $\lambda^*(B)$ exists for each $B \subseteq X$.

The function λ^* just defined has the advantage that it is defined for arbitrary subsets of X rather than just those sets in $\mathcal{C}$. However, it is not a measure in the true sense of the word since it can be seen from the following example that λ^* is not necessarily even finitely additive.

Example 2.1.22

Let X be any non-empty set and let $\mathcal{C} = \{\phi, X\}$. $\mathcal{C}$ is an algebra of sets and the set function λ defined by $\lambda(\phi) = 0$ and $\lambda(X) = 1$ is a measure on $\mathcal{C}$. Let A be any non-empty subset of X such that $A \neq X$. It is easily verified that

$$1 = \lambda^*(X) = \lambda^*[A \cup (X - A)]$$

but that

$$\lambda^*(A) + \lambda^*(X - A) = 2.$$

λ^* does however possess the following properties.

**Theorem* 2.1.23

Let X be a set, $\mathcal{C}$ an algebra of subsets of X, λ a measure on $\mathcal{C}$ and λ^* the outer measure on the power set of X generated by λ. Then

i) $\lambda^*(\phi) = 0$;

ii) $\lambda^*(B) \geq 0$ for $B \subseteq X$;

iii) If $A \subseteq B$, then $\lambda^*(A) \leq \lambda^*(B)$;

iv) If $A \in \mathcal{C}$, then $\lambda^*(A) = \lambda(A)$;

v) If (A_n) is a sequence of subsets of X then

$$\lambda^*\left(\bigcup_{n=1}^{\infty} A_n\right) \leq \sum_{n=1}^{\infty} \lambda^*(A_n).$$

Note 2.1.24

Property v) is often referred to as the *countable subadditive property* of outer measure. This property will be of interest with respect to probability measures. The reader is referred to Chapter Three.

We have now at our disposal a sigma algebra of subsets of X which contains $\mathcal{C}$; namely the power set of X, ρ_X. However, we do not have a measure defined on ρ_X but only an outer measure λ^*. We shall now restrict the domain of λ^* to a smaller sigma algebra of subsets of X, $\mathcal{C}^*$, such that $\mathcal{C} \subseteq \mathcal{C}^*$ and λ^* restricted to $\mathcal{C}^*$ is a measure.

Definition 2.1.25

A subset A of X is said to be λ^* *measurable* if

$$\lambda^*(Y) = \lambda^*(Y \cap A) + \lambda^*(Y - A)$$

for all subsets Y of X. We shall denote the collection of all λ^* measurable subsets of X by $\mathcal{C}^*$.

The following theorem provides the basis for the technique of Lebesgue.

***Theorem 2.1.26**

(Caratheodory Extension Theorem)

The ordered triple $(X, \mathcal{C}^*, \lambda^*)$ is a measure space such that

$$\mathcal{C} \subseteq \mathcal{C}^* \quad \text{and}$$

$$\lambda^*(A) = \lambda(A) \quad \text{for } A \in \mathcal{C}.$$

To obtain the usual Lebesgue measure on the real line we need only apply the above extension procedure to algebra F with measure ℓ defined on F. In this way we obtain a measure space $(\mathcal{R}, F^*, \ell^*)$ with the desired properties namely that F^* contains every finite interval I and that $\ell^*(I) = \text{length } I$.

Note 2.1.27

It can be shown that the smallest σ algebra containing F is exactly the collection of Borel sets defined in *Example* (2.1.3). Since $F \subseteq F^*$ we must have by *Definition* (1.1.25) that every Borel set B is also a Lebesgue measurable set. Hence we can restrict Lebesgue measure ℓ^* to the collection β of Borel sets and obtain a measure space $(\mathcal{R}, \beta, \ell^*|\beta)$. The restriction of Lebesgue measure to Borel sets is sometimes referred to as Borel measure.

We shall have occasion in later chapters to consider what will be referred to as Borel subsets of the open unit interval (0, 1). By this phrase we shall mean the following:

Definition 2.1.28

Consider the open unit interval (0, 1). By the *Borel subsets of* (0, 1) we shall mean all sets of the form $(0, 1) \cap B$ where B is an element of β. We shall denote this collection by β^*.

Note 2.1.29

It is possible to define the Borel sets of (0, 1) in a manner similar to that used in *Example* (2.1.3). Namely the Borel sets of (0, 1) is the σ algebra generated by the collection K of all sets of the form $(0, 1) \cap U$ where U denotes an open set in R with respect to the usual Euclidean topology on the real line. Note that K is simply the relative topology on (0, 1). (See *Definition* (1.2.9)).

**Theorem* 2.1.30

Let (X, C^*, λ^*) be the measure space obtained by the Caratheodory extension procedure. C^* is complete in the sense that if $B \in C^*$, $\lambda^*(B) = 0$ and $A \subseteq B$, then $A \in C^*$ and $\lambda^*(A) = 0$.

Note 2.1.31

The concept of completeness will play an important role in the theory of probability and will be referred to in later chapters.

Theorem 2.1.32

If A is a countable subset of R then A is a Borel set and A has Lebesgue measure zero.

Definition 2.1.33

Let $(X, \mathcal{C})$ and $(Y, \mathcal{K})$ be measurable spaces and let $f: X \to Y$. f is said to be *measurable* if and only if given any $F \in \mathcal{K}$, $f^{-1}(F) \in \mathcal{C}$.

There is a special case of the above definition which is of interest in the study of probability theory. This is the case where the function f actually maps from a measure space which we shall denote $(\Omega, \mathcal{F}, P)$ into the measurable space (R, β) where β is the collection of Borel sets. In this case we shall have at hand the tools required to define what is known as a random variable in probabilistic terms.

Definition 2.1.34

Let $(X, \mathcal{C})$ be a measurable space and let f be a real valued function defined on X. f is said to be *measurable* if $f^{-1}(E) \in \mathcal{C}$ for every Borel set E.

The following series of theorems help to provide a characterization of the term measurable function in the case of a real valued function f from a measurable space. *Theorem* (2.1.38) provides essentially an equivalent form for *Definition* (2.1.34) which will be useful in later work.

Theorem 2.1.35

Let $f: X \to Y$. Let $\mathcal{C}$ be a σ algebra of subsets of X and let $\mathcal{K} = \{E \subseteq Y : f^{-1}(E) \in \mathcal{C}\}$. Then $\mathcal{K}$ is a σ algebra of subsets of Y.

Proof: $f^{-1}(\phi) = \phi$ by *Theorem* (1.2.13). Since $\mathcal{C}$ is a σ algebra, $\phi \in \mathcal{C}$ and ϕ is therefore an element of $\mathcal{K}$. Now let $E \in \mathcal{K}$ and consider $f^{-1}(Y - E)$. By *Theorem* (1.2.13), $f^{-1}(Y - E) = f^{-1}(Y) - f^{-1}(E)$. Hence by *Theorem* (1.2.13), $f^{-1}(Y - E) = X - f^{-1}(E)$. Since $E \in \mathcal{K}$,

$f^{-1}(E) \in \mathcal{C}$ by definition. Since $\mathcal{C}$ is a σ algebra, $X - f^{-1}(E) \in \mathcal{C}$ by *Definition* (1.1.22). Hence $f^{-1}(Y - E) \in \mathcal{C}$ which implies that $Y - E \in \mathcal{K}$. Now let $\{E_\gamma : \gamma \in \Gamma\}$ be a countable collection of elements of $\mathcal{K}$. Consider $f^{-1}(\bigcup_{\gamma\in\Gamma} E_\gamma)$. By *Theorem* (1.2.13) $f^{-1}(\bigcup_{\gamma\in\Gamma} E_\gamma) = \bigcup_{\gamma\in\Gamma} f^{-1}(E_\gamma)$. By definition of $\mathcal{K}$, for each $\gamma \in \Gamma$ we have $f^{-1}(E_\gamma) \in \mathcal{C}$. Since $\mathcal{C}$ is a σ algebra $\bigcup_{\gamma\in\Gamma} f^{-1}(E_\gamma) \in \mathcal{C}$ implying that $f^{-1}(\bigcup_{\gamma\in\Gamma} E_\gamma) \in \mathcal{C}$. This is sufficient for $\bigcup_{\gamma\in\Gamma} E_\gamma$ to be in $\mathcal{K}$ as was required.

Theorem 2.1.36

Let $(X, \mathcal{C})$ be a measurable space and let $f : X \to Y$. Let $\mathcal{A}$ be a collection of subsets of Y such that $f^{-1}(E) \in \mathcal{C}$ for every $E \in \mathcal{A}$. Then $f^{-1}(F) \in \mathcal{C}$ for any set F which belongs to the σ algebra generated by $\mathcal{A}$.

Proof: Let $\mathcal{K} = \{E \subseteq Y : f^{-1}(E) \in \mathcal{C}\}$. Then $\mathcal{A} \subseteq \mathcal{K}$ by definition and $\mathcal{K}$ is a σ algebra containing $\mathcal{A}$ by *Theorem* (2.1.35). Let $\mathcal{G}$ be the σ algebra generated by $\mathcal{A}$. Then $\mathcal{A} \subseteq \mathcal{G} \subseteq \mathcal{K}$ by *Definition* (1.1.25). Let $F \in \mathcal{G}$. Then also $F \in \mathcal{K}$. By definition of $\mathcal{K}$, $f^{-1}(F) \in \mathcal{C}$ as was desired.

Theorem 2.1.37

Let $(X, \mathcal{C})$ be a measurable space and let f be a real valued function defined on X. f is measurable if and only if $f^{-1}(-\infty,\alpha) \in \mathcal{C}$ for every real number α.

Proof: Assume that f is measurable. Choose α an arbitrary but fixed real number. Let N_1 be the smallest natural number such that $-N_1 < \alpha$. Note that for each natural number $n \geq N_1$ $(-n, \alpha)$ is a Borel set by *Example* (2.1.3). Note also that

$$(-\infty, \alpha) = \bigcup_{n=N_1}^{\infty} (-n, \alpha).$$

Since the Borel sets form a σ algebra, $(-\infty, \alpha)$ is also a Borel set by the countable additivity property of a σ algebra. Since f is measurable, $f^{-1}(-\infty, \alpha) \varepsilon\, \mathcal{C}$ by *Definition* (2.1.34). Now assume that $f^{-1}(-\infty, \alpha) \varepsilon\, \mathcal{C}$ for each real number α. Consider the open interval (a, b) and write $(a, b) = (-\infty, b) \cap (a, \infty)$. Now $f^{-1}(a, b) = f^{-1}(-\infty, b) \cap f^{-1}(a,\infty)$ by *Theorem* (1.2.13). We wish to show that $f^{-1}(a, b) \varepsilon\, \mathcal{C}$. By assumption $f^{-1}(-\infty, b) \varepsilon\, \mathcal{C}$ so it remains only to argue that $f^{-1}(a, \infty) \varepsilon\, \mathcal{C}$. Write

$$(a, \infty) = \bigcup_{n=1}^{\infty} [a + \frac{1}{n}, \infty). \quad \text{Hence}$$

$$f^{-1}(a, \infty) = f^{-1}(\bigcup_{n=1}^{\infty} [a + \frac{1}{n}, \infty)) = \bigcup_{n=1}^{\infty} f^{-1}[a + \frac{1}{n}, \infty)$$

by *Theorem* (1.2.13). The problem thus becomes to argue that for each n, $f^{-1}[a + \frac{1}{n}, \infty) \varepsilon\, \mathcal{C}$. To see this note that for each n

$$(-\infty, a + \frac{1}{n})' = [a + \frac{1}{n}, \infty).$$

Hence $f^{-1}(-\infty, a + \frac{1}{n})' = f^{-1}[a + \frac{1}{n}, \infty)$.

However $f^{-1}(-\infty, a + \frac{1}{n})' = [f^{-1}(-\infty, a + \frac{1}{n})]'$ by *Theorem* (1.2.13). Since $f^{-1}(-\infty, a + \frac{1}{n}) \in \mathcal{C}$ by assumption and $\mathcal{C}$ is a σ algebra, $[f^{-1}(-\infty, a + \frac{1}{n})]' \in \mathcal{C}$ for each n. Thus we obtain that for each n, $f^{-1}[a + \frac{1}{n}, \infty) \in \mathcal{C}$ as was desired. Now let $\mathcal{A} = \{(a, b) : a, b \text{ real}\}$. We have just shown that $f^{-1}(a, b) \in \mathcal{C}$ and hence $\mathcal{A}$ satisfies the condition of *Theorem* (2.1.36) and we may conclude that $f^{-1}(F) \in \mathcal{C}$ for any set F which belongs to the σ algebra generated by $\mathcal{A}$. By *Example* (2.1.3) the σ algebra generated by $\mathcal{A}$ is precisely the Borel algebra. We have thus shown that for any Borel set F, $f^{-1}(F) \in \mathcal{C}$ which by *Definition* (2.1.34) implies that f is measurable and the proof is complete.

The following theorem provides several other equivalent forms for *Definition* (2.1.34) which will be of interest in Chapter Four.

Theorem 2.1.38

Let $(X, \mathcal{C})$ be a measurable space and let f be a real valued function defined on X. Then the following statements are equivalent:

i) f is measurable;

ii) $f^{-1}(-\infty, \alpha) \in \mathcal{C}$ for $\alpha \in \mathcal{R}$;

iii) $f^{-1}[\alpha, \infty) \in \mathcal{C}$ for $\alpha \in \mathcal{R}$;

iv) $f^{-1}(\alpha, \infty) \in \mathcal{C}$ for $\alpha \in \mathcal{R}$;

v) $f^{-1}(-\infty, \alpha] \in \mathcal{C}$ for $\alpha \in \mathcal{R}$.

Proof: i) $\Longleftrightarrow$ ii) is a direct restatement of *Theorem* (2.1.37). ii) $\Longleftrightarrow$ iii) follows from the fact that $(-\infty, \alpha) = [\alpha, \infty)'$ and hence

$f^{-1}(-\infty, \alpha) = f^{-1}[\alpha, \infty)' = [f^{-1}[\alpha, \infty)]'$ by *Theorem* (1.2.13). Thus if $f^{-1}(-\infty, \alpha) \varepsilon C$, then $[f^{-1}(-\infty, \alpha)]' = f^{-1}[\alpha, \infty) \varepsilon C$ since C is a σ algebra. Similarly $f^{-1}[\alpha, \infty) \varepsilon C$ implies that $[f^{-1}[\alpha, \infty)]' = f^{-1}(-\infty, \alpha) \varepsilon C$ and we obtain ii) <—> iii). The argument for iv) <—> v) is identical to the above. To show that iii) $\rightarrow$ iv), note that

$$(\alpha, \infty) = \bigcup_{n=1}^{\infty} [\alpha + \frac{1}{n}, \infty). \quad \text{Thus}$$

$$f^{-1}(\alpha, \infty) = f^{-1}(\bigcup_{n=1}^{\infty} [\alpha + \frac{1}{n}, \infty))$$

$$= \bigcup_{n=1}^{\infty} f^{-1}[\alpha + \frac{1}{n}, \infty) \quad \text{by } \textit{Theorem } (1.2.13).$$

By hypothesis, for each n $f^{-1}[\alpha + \frac{1}{n}, \infty) \varepsilon C$. Since C is a σ algebra, $\bigcup_{n=1}^{\infty} f^{-1}[\alpha + \frac{1}{n}, \infty) \varepsilon C$ as was desired.

To show that iv) $\rightarrow$ iii), note that

$$[\alpha, \infty) = \bigcap_{n=1}^{\infty} (\alpha - \frac{1}{n}, \infty). \quad \text{Thus}$$

$$f^{-1}[\alpha, \infty) = f^{-1}(\bigcap_{n=1}^{\infty} (\alpha - \frac{1}{n}, \infty))$$

$$= \bigcap_{n=1}^{\infty} f^{-1}(\alpha - \frac{1}{n}, \infty) \quad \text{by } \textit{Theorem } (1.2.13).$$

By hypothesis, for each n $f^{-1}(\alpha - \frac{1}{n}, \infty) \varepsilon C$. Since C is a σ algebra, $\bigcap_{n=1}^{\infty} f^{-1}(\alpha - \frac{1}{n}, \infty) \varepsilon C$ as was desired and the proof is complete.

Corollary 2.1.39

Let (X, C) be a measurable space and let f be a

real valued function defined on X. Then

$$\{x : f(x) = \alpha\} \in \mathcal{C}$$

for each $\alpha \in R$.

Note 2.1.40

If we take the space $(X, \mathcal{C})$ of *Definition* (2.1.34) to be the measurable space (R, β) where R is the set of real numbers and β the collection of Borel sets then we refer to f as being Borel measurable.

Theorem. 2.1.41

Any open set in the Euclidean topology E on the real line is also a Borel set.

Proof: By *Theorem* (1.2.11), (R, E) is second countable and in particular by *Theorem* (1.2.8) $B_1 = \{(d_1, d_2) : d_1$ and d_2 are rational$\}$ is a countable base for E. Let U be an open set. By *Definition* (1.2.3) U can be expressed as the union of a collection of elements of B_1. Since B_1 is countable, U can be expressed as a countable union of open intervals. By *Example* (2.1.3) each open interval is a Borel set. Since the collection of Borel sets forms a σ algebra any countable union of Borel sets is a Borel set. In particular U is a Borel set as was to be shown.

Theorem 2.1.42

Let (R, β) be the measurable space of *Note* (2.1.40) and let $f : R \to R$ where we assume the usual Euclidean topology on R. If f is continuous, then f is Borel measurable.

Proof: To show that f is Borel measurable it is necessary to show that $f^{-1}(E) \varepsilon \beta$ for each Borel set E. By *Theorem* (2.1.38) it is sufficient to show that for each real number α, $f^{-1}(-\infty, \alpha) \varepsilon \beta$. However since $(-\infty, \alpha)$ is open in the Euclidean topology and f is continuous, $f^{-1}(-\infty, \alpha)$ is open by *Theorem* (1.2.16). By *Theorem* (2.1.41) $f^{-1}(-\infty, \alpha)$ is a Borel set as was to be shown.

Theorem 2.1.43

Let (R, β) be the measurable space of *Note* (2.1.40) and let $f : R \to R$. Any monotone function f (either increasing in the sense that if $x_1 \leq x_2$ then $f(x_1) \leq f(x_2)$ or decreasing in the sense that if $x_1 \leq x_2$ then $f(x_1) \geq f(x_2)$) is Borel measurable.

Proof: We shall consider the case of a decreasing function f. The argument for the increasing case is analogous. Let $\alpha \varepsilon R$. If $f(x) = \alpha$ for some interval of the form $(-\infty, c)$ or $(-\infty, c]$ c real or if f is a step function then the proof is trivial. Let us assume therefore that f is not a step function and $f(x) > \alpha$ for some x.

Assume that there exists a real number q such that $f(q) = \alpha$. Let $S = \{x : f(x) = \alpha\}$. If there exists no real number x such that $f(x) > \alpha$, then $f^{-1}(\alpha, \infty) = \phi$ is a Borel set as desired. If there exists a real number t such that $f(t) > \alpha$, then t is a lower bound for S since if not $t \geq z$ for some $z \varepsilon S$. This implies that $f(t) \leq f(z) = \alpha$ which is a contradiction. Thus S is a non-empty set of real numbers which is bounded below and hence has a greatest lower bound which we shall denote by a. Note that since f is decreasing S either consists of a single point or else S is an interval of real numbers closed on the left. In either case $a \in S$. We claim that $f^{-1}(\alpha, \infty) = (-\infty, a)$. To show this let $x \in f^{-1}(\alpha, \infty)$. Then $f(x) > \alpha$ and as argued previously, x is a lower bound for S implying that $x \leq a$. It is evident that $x \neq a$, since $f(a) = \alpha$. Hence $x < a$ and we obtain that $f^{-1}(\alpha, \infty) \subseteq (-\infty, a)$. To show that if $x \in (-\infty, a)$ then $f(x) > \alpha$, assume that $x < a$ but $f(x) \leq \alpha$. Since $x < a$, $f(x) \geq f(a) = \alpha$. Hence we must have that $f(x) = \alpha$. This implies that $x \in S$ and that $a \leq x$ which is a contradiction. We may thus conclude that $(-\infty, \alpha) \subseteq f^{-1}(\alpha, \infty)$. Hence $f^{-1}(\alpha, \infty) = (-\infty, a)$ as was to be shown. Since $(-\infty, a)$ is a Borel set the proof is complete for the case at hand. It must still be argued that the theorem is true if there exists no real number q such that $f(q) = \alpha$. If $f(x) > \alpha$ for all real numbers x, then $f^{-1}(\alpha, \infty) = \mathcal{R}$ and if $f(x) < \alpha$ for all real numbers x then $f^{-1}(\alpha, \infty) = \phi$. In either of these cases, $f^{-1}(\alpha, \infty)$ is a Borel set as desired. Assume that there exist real numbers x_1 and x_2 such that $f(x_1) > \alpha$ and $f(x_2) < \alpha$. Consider $S = \{x : f(x) < \alpha\}$. Then $S \neq \phi$ and S is bounded below by x_1. Hence S has a greatest lower bound which we denote by a. If $a \notin S$, then $f(a) > \alpha$. We claim that in this case, $f^{-1}(\alpha, \infty) = (-\infty, a] = \bigcap_{n=1}^{\infty} (-\infty, a + \frac{1}{n})$ which is a Borel set. To show this, let $x \in f^{-1}(\alpha, \infty)$. Then $f(x) > \alpha$ and as was argued previously, x is a lower bound for S implying that $x \leq a$ and hence that $f^{-1}(\alpha, \infty) \subseteq (-\infty, a]$. To show that if $x \leq a$, then $f(x) > \alpha$ assume that $x \leq a$ but $f(x) < \alpha$. Note that $x \leq a$ implies $f(x) \geq f(a) > \alpha$

which is a contradiction. We thus obtain that $(-\infty, a] \subseteq f^{-1}(\alpha, \infty)$ which by *Definition* (1.1.6) implies that $f^{-1}(\alpha, \infty) = (-\infty, a]$ as was desired. If $a \in S$ then $f(a) < \alpha$. We claim that in this case $f^{-1}(\alpha, \infty) = (-\infty, a)$. Let $x \in f^{-1}(\alpha, \infty)$. Then once again $f(x) > \alpha$ and x is a lower bound for S implying that $x \leq a$. $x \neq a$ since this would imply that $f(x) = f(a)$ and that $f(x) > \alpha$ and $f(x) < \alpha$ simultaneously. Hence $f^{-1}(\alpha, \infty) \subseteq (-\infty, a)$. To show that if $x < a$, $f(x) > \alpha$ assume that $f(x) < \alpha$. Then $x \in S$ implying that $a \leq x$. This last result is the contrapositive of the desired result and hence we may conclude that $(-\infty, a) \subseteq f^{-1}(\alpha, \infty)$. Thus $f^{-1}(\alpha, \infty) = (-\infty, a)$ as was desired and the proof is complete.

Note that the emphasis thus far has been placed upon *real valued* measurable functions defined on some measurable space $(X, \mathcal{C})$. (See *Definition* (2.1.34), *Theorem* (2.1.37), *Theorem* (2.1.38)). It will be necessary in section two to extend this concept to functions mapping X into $\overline{R}$ the extended real numbers (see *Note* (2.1.4)). To this end we make the following definition:

Definition 2.1.44

Let $(X, \mathcal{C})$ be a measurable space and let $f : X \to \overline{R}$. f is said to be *measurable* if $\{x : f(x) > \alpha\} \in \mathcal{C}$ for each real number α.

The only difference of course in the above definition and that given in *Theorem* (2.1.38) is that f is now allowed to take on the "values" $\pm\infty$ implying that $\{x : f(x) = +\infty\} \subseteq \{x : f(x) > \alpha\}$ for each $\alpha \in R$.

We have concentrated on functions which map either into the real numbers or the extended real numbers. We shall have occasion in Chapter Six to consider functions which map into Euclidean n-space. We summarize here some information concerning such functions which will be pertinent to our study of n-dimensional random variables.

Definition 2.1.45

If $a = (a_1, a_2, \ldots, a_n)$ and $b = (b_1, b_2, \ldots, b_n)$ are elements of $\mathcal{R}^n$ with $a_i < b_i$ for $i = 1, 2, \ldots, n$ then by the *half open interval* $(a, b]$ in $\mathcal{R}^n$ we shall mean

$$\{(x_1, x_2, \ldots, x_n) : a_i < x_i \leq b_i, \ i = 1, 2, \ldots, n\}.$$

Definition 2.1.46

The *Borel sets of* $\mathcal{R}^n$, denoted β^n are defined to be the σ algebra generated by all half open intervals $(a, b]$ in $\mathcal{R}^n$.

Note 2.1.47

It can be shown, Taylor [9], that the Borel sets are also generated by the elements of $\mathcal{E}^n$.

Note 2.1.48

When $n = 1$, $\beta^n = \beta$ of *Example* (2.1.3).

Definition 2.1.49

Let $(X, \mathcal{C})$ be a measurable space and let $f : X \to \mathcal{R}^n$. f is said to be *measurable* if $f^{-1}(E) \in \mathcal{C}$ for every Borel set E of $\mathcal{R}^n$.

Note 2.1.50

The preceeding definition is a natural extension of *Definition* (2.1.34). As was the case previously it is possible to rephrase *Definition* (2.1.49) to obtain an equivalent definition as follows: "Let $(X, \mathcal{C})$ be a measurable space and let $f : X \to R^n$. f is measurable if and only if $f^{-1}(-\infty, b] \varepsilon \mathcal{C}$ for any point $b \varepsilon R^n$." Note also that if we take $(X, \mathcal{C}) = (R^m, \beta^m)$ and let $f : X \to R^n$ be measurable then we say that f is Borel measurable. This is a natural extension of *Note* (2.1.40).

Theorem 2.1.51

Let $(X, \mathcal{C})$ be a measurable space and let $f : X \to R^n$ be such that $f = (f_1, f_2, . . . , f_n)$. (That is, we define $f(x) = (f_1(x), f_2(x), . . . , f_n(x))$ where for each $i = 1, 2, . . n$, f_i is a function from X into R). f is measurable if and only if f_i is measurable for each $i = 1, 2, ... n$.

Proof: Assume that for each i, f_i is measurable. Let $b = (b_1, b_2, . . . ,b_n) \varepsilon R^n$. Consider

$$f^{-1}(-\infty, b] = \bigcap_{i=1}^{n} f_i^{-1}(-\infty, b_i]. \quad \text{Since}$$

for each i f_i is measurable, by *Theorem* (2.1.38) $f_i^{-1}(-\infty, b_i] \varepsilon \mathcal{C}$ $i = 1, 2, ... n$. Since $\mathcal{C}$ is a σ algebra $\bigcap_{i=1}^{n} f_i^{-1}(-\infty, b_i] \varepsilon \mathcal{C}$. Thus f is measurable by *Note* (2.1.50). Now assume that f is measurable. Consider $f_i^{-1}(-\infty, \alpha]$ for α real. Define a sequence of sets (E_k) by

$$E_k = \{x \in X : f_1(x) \leq k,\ f_2(x) \leq k,\ \ldots\ , f_{i-1}(x) \leq k,$$
$$f_i(x) \leq \alpha,\ f_{i+1}(x) \leq k,\ \ldots\ , f_n(x) \leq k\}.$$

Note that since $(k, k, \ldots\ k, \alpha, k, \ldots\ , k) \in \mathcal{R}^n$ and since f is measurable for each $k = 1, 2, \ldots$ we have by *Note* (2.1.50) that $E_k \in \mathcal{C}$ implying that $\bigcup_{k=1}^{\infty} E_k \in \mathcal{C}$ since $\mathcal{C}$ is a σ algebra. To complete the proof, note that

$$f_i^{-1}(-\infty, \alpha] = \bigcup_{k=1}^{\infty} E_k$$

and apply *Theorem* (2.1.38).

The following exercises will give the reader some experience in dealing with *Definition* (2.1.7) and will be useful later in connection with our development of the Lebesgue integral.

2.1 *EXERCISES*

1. Let $(X, \mathcal{C}, \lambda)$ be a measure space and let A be an arbitrary but fixed element of $\mathcal{C}$. Let $E \in \mathcal{C}$. Define a function γ on $\mathcal{C}$ by $\gamma(E) = \lambda(A \cap E)$. Then γ is a measure on $\mathcal{C}$.
2. Let $\lambda_1, \lambda_2, \ldots\ , \lambda_n$ be measures with respect to the measurable space $(X, \mathcal{C})$. Let $a_1, a_2, \ldots\ , a_n$ be non-negative real numbers and define a function γ on $\mathcal{C}$ by

$$\gamma(E) = \sum_{i=1}^{n} a_i \lambda_i(E)$$

for $E \in \mathcal{C}$. Then γ is a measure on $\mathcal{C}$.

3. Prove that the set of all irrational numbers in $[0, 1]$ is Borel measurable and find its measure.
4. Let $(X, \mathcal{C}, \lambda)$ be a measure space. Let f_1 be a continuous function of a real variable and $f_2 : X \to \mathcal{R}$ be measurable. Prove that $f_1(f_2(x))$ is measurable.

2.2 *A BRIEF INTRODUCTION TO THE LEBESGUE INTEGRAL*

We present in this section an outline of the technique used to define the Lebesgue integral of real valued functions f defined with respect to an underlying measure space $(X, \mathcal{C}, \lambda)$. In order to do so we shall first define the integral for simple non-negative real valued functions ψ measurable with respect to $(X, \mathcal{C}, \lambda)$. We shall then extend this concept to include non-negative extended real valued measurable functions and finally real valued measurable functions which are not necessarily non-negative. Our main sources of reference are Bartle [2] and Royden [7]. The concept of the Lebesgue integral plays a major role in the study of probability theory as it underlies the idea of expectation of a random variable which is the topic of Chapter Four.

We shall be assuming throughout this section the existence of an underlying measure space $(X, \mathcal{C}, \lambda)$. When we say that f is a real valued measurable function or an extended real valued measurable function defined on X we shall mean that f is measurable in the sense of *Definition* (2.1.34) or *Definition* (2.1.44) respectively.

Definition 2.2.1

Let $\psi: X \to R$. ψ is said to be *simple* if its range is a finite set.

Definition 2.2.2

Let $A \subseteq X$. Let $\chi_A : X \to \{0, 1\}$ be defined by

$$\chi_A(x) = \begin{cases} 1 & \text{if } x \in A \\ 0 & \text{if } x \notin A \end{cases}$$

χ_A is called the *indicator function* of the set A.

Lemma 2.2.3

Let ψ be a simple function defined on X with range $\{a_1, a_2, a_3, \ldots, a_n\}$. Let $\{E_i : i = 1, 2, 3, \ldots, n\}$ be the collection of subsets of X such that $E_i = \psi^{-1}\{a_i\}$. The collection $\{E_i : i = 1, 2, \ldots, n\}$ is a partition of X.

Proof: Let $x \in X$. Since ψ has as its domain the set X, $\psi(x) = a_i$ for some i. Hence $x \in \psi^{-1}\{a_i\} = E_i$. By *Definition* (1.1.11) $x \in \bigcup_{i=1}^{n} E_i$. Hence $X \subseteq \bigcup_{i=1}^{n} E_i$. Since for each i $E_i \subseteq X$ it is evident that $\bigcup_{i=1}^{n} E_i \subseteq X$ and hence by *Definition* (1.1.6), $X = \bigcup_{i=1}^{n} E_i$.

Assume that there exists an element $x \in X$ such that $x \in E_i \cap E_j$ for $i \neq j$. Then $\psi(x) = a_i$ and $\psi(x) = a_j$. Since ψ is a function, $a_i = a_j$ $i \neq j$ which is a contradiction of the choice of the set $\{a_1, a_2, \ldots, a_n\}$.

Hence $E_i \cap E_j = \phi$, $i \neq j$ and $\{E_i : i = 1, 2, 3, \ldots, n\}$ is a partition of X.

Definition 2.2.4

Let ψ be a simple function on X with range $\{a_1, a_2, \ldots, a_n\}$. Let $\{E_i : i = 1, 2, 3, \ldots, n\}$ be chosen as in *Lemma* (2.2.3). Then

$$\psi = \sum_{i=1}^{n} a_i \chi_{E_i}$$

is called the *standard representation for* ψ.

There are many representations for ψ as a linear combination of indicator functions. However there is only one with the properties obtained in *Definition* (2.2.4) namely that the elements $a_1, a_2, \ldots, a_n$ are distinct and the collection $\{E_i : i = 1, 2, 3, \ldots, n\}$ is a partition of X.

Example 2.2.5

Let $X = N$ and let $\mathcal{C} = \rho_X = \rho_N$. Define the simple function ψ by

$$\psi(x) = \begin{cases} 2 & \text{if } x \text{ is even} \\ 3 & \text{if } x \text{ is odd.} \end{cases}$$

The standard representation for ψ is given by

$$\psi = 2\chi_{E_1} + 3\chi_{E_2}$$

where $E_1 = \{\text{even natural numbers}\}$

E_2 = {odd natural numbers}.

We shall denote by $M^+ = M^+(X, \mathcal{C})$ the set of all non-negative functions $f : X \to \overline{R}$ measurable with respect to $(X, \mathcal{C}, \lambda)$ and we shall denote by $M = M(X, \mathcal{C})$ the set of all functions $f : X \to \overline{R}$ measurable with respect to $(X, \mathcal{C}, \lambda)$.

Note 2.2.6

Neither of the sets M or M^+ defined above is empty.

Proof: Since $\mathcal{C}$ is a σ algebra, $X \in \mathcal{C}$. Consider χ_X. χ_X is non-negative since it takes on only the value 1. Let B be any Borel subset of $\mathcal{R}$. Either $1 \in B$ in which case $\chi_X^{-1}(B) = X \in \mathcal{C}$ or $1 \notin B$ in which case $\chi_X^{-1}(B) = \phi \in \mathcal{C}$. In either case, $\chi_X^{-1}(B) \in \mathcal{C}$ and hence by *Definition* (2.1.34) χ_X is measurable. Thus $\chi_X \in M$ and also $\chi_X \in M^+$.

Definition 2.2.7

If ψ is a simple function in $M^+(X, \mathcal{C})$ with standard representation as in *Definition* (2.2.4) then we define the integral of ψ with respect to λ, denoted $\int\psi d\lambda$, by

$$\int\psi d\lambda = \sum_{i=1}^{n} a_i \lambda(E_i).$$

Example 2.2.8

i) Let $(X, \mathcal{C})$ be as defined in *Example* (2.2.5). Define λ by

$$\lambda\{n\} = \frac{1}{2^n}, \quad \lambda(A) = \sum_{n \in A} \lambda\{n\}$$

for $A \neq \phi$, $\lambda(\phi) = 0$.

$$\int \psi d\lambda = 2 \sum_{i=1}^{\infty} \frac{1}{2^{2i}} + 3 \sum_{i=1}^{\infty} \frac{1}{2^{2i-1}} .$$

ii) Let $(X, \mathcal{C}, \lambda)$ be a general measure space.

$$\int d\lambda = \lambda(X).$$

Verification:

i) $\int \psi d\lambda = 2\,\lambda(E_1) + 3\,\lambda(E_2)$ where

E_1 = {even natural numbers}

E_2 = {odd natural numbers}.

By definition of λ, $\lambda(E_1) = \sum_{i=1}^{\infty} \frac{1}{2^{2i}}$ and

$\lambda(E_2) = \sum_{i=1}^{\infty} \frac{1}{2^{2i-1}}$.

ii) $1 = 1 \cdot \chi_{E_1} + 0.\ \chi_{E_2}$ where

$E_1 = X$ and $E_2 = \phi$. Thus

$$\begin{aligned} \int d\lambda &= 1 \cdot \lambda(E_1) + 0 \cdot \lambda\ (\phi) \\ &= \lambda(E_1) + 0 \\ &= \lambda(X). \end{aligned}$$

Example 2.2.9

Let $(X, \mathcal{C}, \lambda) = ((0, 1), \beta^*, \ell^*|_{\beta^*})$ where $(0, 1)$ is the open unit interval, β^* is the collection of Borel subsets of $(0, 1)$ and $\ell^*|_{\beta^*}$ is the Lebesgue measure on $\mathcal{R}$ restricted to β^*. Define a function ψ mapping $(0, 1) \to \{0, 1\}$ by

$$\psi(x) = \begin{cases} 0 & \text{if } x \in E_1 \\ 1 & \text{if } x \in E_2 \end{cases}$$

where

$$E_1 = (0, 1) \cap \{\text{rational numbers}\}$$

$$E_2 = (0, 1) \cap \{\text{irrational numbers}\}.$$

Then $\psi \in M^+$ and $\int \psi d\lambda = 1$.

Verification:

Note first that for any singleton set $\{a\} \subseteq R$ $\{a\} = [(-\infty, a) \cup (a, \infty)]'$. Since

$(-\infty, a) = \bigcup_{n=1}^{\infty} (-n, a)$ and

$(a, \infty) = \bigcup_{n=1}^{\infty} (a, n)$ by

Example (2.1.3) $(-\infty, a)$ and (a, ∞) are each Borel sets. Since the Borel sets form a σ algebra, $(-\infty, a) \cup (a, \infty)$ and $[(-\infty, a) \cup (a, \infty)]'$ are Borel sets. Thus any singleton set $\{a\} \subseteq R$ is a Borel set implying that any countable subset of R is a Borel set. Since the rational numbers are countable they form a Borel set. By *Definition* (2.1.28), $E_1 \in \beta^*$. Since β^* is a σ algebra and $E_2 = E_1'$, $E_2 \in \beta^*$. Let B be any Borel subset of R. Note that if $\{0, 1\} \subseteq B$, $\psi^{-1}(B) = (0, 1) \in \beta^*$. If $0 \in B$ and $1 \notin B$, $\psi^{-1}(B) = E_1 \in \beta^*$. If $1 \in B$ and $0 \notin B$, $\psi^{-1}(B) = E_2 \in \beta^*$. If $0 \notin B$ and $1 \notin B$, $\psi^{-1}(B) = \phi \in \beta^*$. In any case $\psi^{-1}(B) \in \beta^*$ implying that ψ is measurable with respect to $((0, 1), \beta^*, \ell^*|_{\beta^*})$. ψ is obviously non-negative and simple. Thus

$$\int \psi d\lambda = 0 \cdot \lambda(E_1) + 1 \cdot \lambda(E_2).$$

Since $\lambda = \ell^*|_{\beta^*}$, by *Theorem* (2.1.32) $\lambda(E_1) = 0$. Since $\lambda(0, 1) = 1$ and $(0, 1) = E_1 \cup E_2$, by *Definition* (2.1.7) $\lambda(E_2) = 1$. We thus obtain that

$$\int \psi d\lambda = 0 \cdot 0 + 1 \cdot 1 = 1.$$

The following lemmas summarize some of the elementary properties of the integral of simple functions in $M^+(X, \mathcal{C})$.

Lemma 2.2.10

Let ψ be a simple function in $M^+(X, \mathcal{C})$. If $c \geq 0$ then $\int c\psi d\lambda = c\int \psi d\lambda$.

Proof: Assume that ψ has standard representation given by

$$\psi = \sum_{i=1}^{n} a_i \, \chi_{E_i}.$$

If $c = 0$, then $c\,\psi$ is the zero function and hence

$$c\,\psi = 0\,\chi_X$$

is the standard representation for $c\,\psi$. Hence

$$\int c\psi d\lambda = 0 \cdot \lambda(X) = 0$$

regardless of whether or not X is of finite measure by *Note* (2.1.4). However

$$0 = 0 \cdot \sum_{i=1}^{n} a_i \lambda(E_i) = 0 \cdot \int \psi d\lambda = c \int \psi d\lambda$$

as was desired. If $c > 0$, then $c\psi$ has standard representation

$$c\psi = \sum_{i=1}^{n} ca_i \chi_{E_i}.$$

Thus $\int c\psi d\lambda = \sum_{i=1}^{n} ca_i \chi_{E_i} = c \sum_{i=1}^{n} a_i \chi_{E_i} = c \int \psi d\lambda$

and the proof is complete.

Lemma 2.2.11

Let ψ and ξ be simple functions in $M^+(X, \mathcal{C})$. Then

$$\int (\psi + \xi) d\lambda = \int \psi d\lambda + \int \xi d\lambda.$$

Proof: Let the standard representations for ψ and ξ be given by

$$\psi = \sum_{i=1}^{n} a_i \chi_{E_i}$$

and

$$\xi = \sum_{j=1}^{m} b_j \chi_{F_j}$$

respectively. Note that the collection $\mathcal{G} = \{E_i \cap F_j : i = 1, 2, \ldots, n; j = 1, 2, \ldots, m\}$ is a partition of X and that $\psi + \xi$ can be expressed as a linear combination of indicator functions in the following manner:

$$\psi + \xi = \sum_{i=1}^{n} \sum_{j=1}^{m} (a_i + b_j) \chi_{E_i \cap F_j}$$

This representation however is not necessarily the standard representation of *Definition* (2.2.4) since the numbers $a_i + b_j$ may not be distinct as required. To obtain the standard representation, let $\{c_k : k = 1, 2, 3, \ldots, r\}$ be the set of distinct elements of $\{a_i + b_j : i = 1, 2, \ldots, n;\ j = 1, 2, \ldots, m\}$ and let G_k be the union of all those sets $E_i \cap F_j$ such that $a_i + b_j = c_k$. Note that since $\mathcal{G}$ is a partition of X and λ is a measure $\lambda(G_k) = \sum_{(k)} \lambda(E_i \cap F_j)$ where the notation indicates that for each k the summation is taken over all i and j such that $a_i + b_j = c_k$. The standard representation for $\psi + \xi$ is now seen to be given by

$$\psi + \xi = \sum_{k=1}^{r} c_k \chi_{G_k} .$$

Thus
$$\begin{aligned}
\int (\psi + \xi)\, d\lambda &= \sum_{k=1}^{r} c_k \lambda(G_k) \\
&= \sum_{k=1}^{r} c_k \sum_{(k)} \lambda(E_i \cap F_j) \\
&= \sum_{k=1}^{r} \sum_{(k)} (a_i + b_j) \lambda(E_i \cap F_j) \\
&= \sum_{i=1}^{n} \sum_{j=1}^{m} (a_i + b_j) \lambda(E_i \cap F_j)
\end{aligned}$$

$$= \sum_{i=1}^{n} \sum_{j=1}^{m} a_i \lambda(E_i \cap F_j) + \sum_{i=1}^{n} \sum_{j=1}^{m} b_j \lambda(E_i \cap F_j).$$

Note that since $\{E_i : i = 1, 2, \ldots, n\}$ and $\{F_j : j = 1, 2, \ldots, m\}$ are both partitions of X,

$\lambda(E_i) = \sum_{j=1}^{m} \lambda(E_i \cap F_j)$ for each i and

$\lambda(F_j) = \sum_{i=1}^{n} \lambda(E_i \cap F_j)$ for each j.

Hence we obtain that

$$\begin{aligned} \int(\psi + \xi)d\lambda &= \sum_{i=1}^{n} a_i \sum_{j=1}^{m} \lambda(E_i \cap F_j) + \sum_{j=1}^{m} b_j \sum_{i=1}^{n} \lambda(E_i \cap F_j) \\ &= \sum_{i=1}^{n} a_i \lambda(E_i) + \sum_{j=1}^{m} b_j \lambda(F_j) \\ &= \int \psi d\lambda + \int \xi d\lambda \end{aligned}$$

and the proof is complete.

Note 2.2.12

The above lemma can be extended to n simple functions in $M^+(X, \mathcal{C})$ by induction.

Corollary 2.2.13

It follows from *Lemma* (2.2.11) that if

$$\psi = \sum_{i=1}^{n} c_i \chi_{E_i}$$

where $c_i \geq 0$, is any representation of ψ, then

$$\int \psi d\lambda = \sum_{i=1}^{n} c_i \lambda(E_i).$$

Lemma 2.2.14

Let ψ be a simple function in $M^+(X, \mathcal{C})$. Let the set function γ be defined for elements E of $\mathcal{C}$ by

$$\gamma(E) = \int \psi \chi_E d\lambda.$$

γ is a measure on $\mathcal{C}$.

Proof: Let ψ have standard representation given by

$$\psi = \sum_{i=1}^{n} a_i \chi_{E_i}.$$

Note that $\psi\chi_E$ can be expressed as a linear combination of indicator functions as follows but that the representation is not necessarily standard:

$$\psi\chi_E = \sum_{i=1}^{n} a_i \chi_{E_i \cap E}.$$

Note also that for each i, $a_i \geq 0$. Thus $\gamma(E) =$ $\int \psi\chi_E d\lambda = \int (\sum_{i=1}^{n} a_i \chi_{E_i \cap E}) d\lambda$. However by *Note* (2.2.12),

$\int (\sum_{i=1}^{n} a_i \chi_{E_i \cap E}) d\lambda = \sum_{i=1}^{n} \int a_i \chi_{E_i \cap E} d\lambda$ and by *Lemma* (2.2.10)

for each i

$\int a_i \, \chi_{E_i \cap E} \, d\lambda = a_i \int \chi_{E_i \cap E} \, d\lambda$. We thus obtain that

$$\gamma(E) = \sum_{i=1}^{n} a_i \int \chi_{E_i \cap E} \, d\lambda.$$

However $\int \chi_{E_i \cap E} \, d\lambda = \lambda(E_i \cap E)$ and hence

$\gamma(E) = \sum_{i=1}^{n} a_i \, \lambda(E_i \cap E)$. Note that by *Exercise* (2.1.1) for each i the map γ_i defined on $\mathcal{C}$ by $\gamma_i(E) = \lambda(E_i \cap E)$ is a measure on $\mathcal{C}$. Hence we can write

$$\gamma(E) = \sum_{i=1}^{n} a_i \, \gamma_i(E)$$

and we have expressed γ as a linear combination of measures on $\mathcal{C}$. By *Exercise* (2.1.2) γ is also a measure on $\mathcal{C}$.

Note that thus far we have defined the integral with respect to λ only for simple functions $\psi \in M^+(X, \mathcal{C})$. We wish to extend this concept to include general functions $f \in M^+(X, \mathcal{C})$. In order to do so we must note the following properties of $\mathcal{R}$ and $\overline{\mathcal{R}}$.

Note 2.2.15

i) Any non-empty set of real numbers which is bounded above has a unique least upper bound or supremum in $\mathcal{R}$;

ii) Every non-empty set of real numbers which is bounded below has a unique greatest lower bound or infimum in $\mathcal{R}$. See Taylor [9];

iii) We consider the infimum of every non-empty set $S \subseteq \overline{R}$ which is not bounded below in the usual sense to be $-\infty$. Thus every non-empty set S of extended real numbers has a unique infimum in $\overline{R}$ which we denote by $\inf S$;

iv) We shall follow the convention that the supremum of a non-empty set $S \subseteq \overline{R}$ which is not bounded above is ∞. Hence every non-empty set S of extended real numbers has a unique supremum in $\overline{R}$ which we shall denote by $\sup S$.

Definition 2.2.16

Let $f \in M^+(X, \mathcal{C})$. The integral of f with respect to λ, denoted $\int f d\lambda$, is defined by

$$\int f d\lambda = \sup \{\int \psi d\lambda : \psi \in M^+(X, \mathcal{C}) \text{ is simple and } 0 \leq \psi(x) \leq f(x) \text{ for } x \in X\}.$$

Note 2.2.17

The integral as defined above always exists and is unique since

$$\{\int \psi d\lambda : \psi \in M^+(X, \mathcal{C}) \text{ is simple and } 0 \leq \psi(x) \leq f(x) \text{ for } x \in X\} \neq \phi.$$

We have in both *Definition* (2.2.7) and *Definition* (2.2.16) defined essentially the "integral of f (or ψ) with respect to λ over the entire set X." It is often desirable to speak of "the integral of f with respect to λ over $E \in \mathcal{C}$" where E is not necessarily the entire set X. With this in mind we make the following definition.

Definition 2.2.18

Let $f \in M^+(X, \mathcal{C})$ and let $E \in \mathcal{C}$. The integral of f with respect to λ over E, denoted $\int_E f d\lambda$, is defined by

$$\int_E f d\lambda = \int f \chi_E d\lambda.$$

Note 2.2.19

The integral defined above exists since $f\chi_E \in M^+(X, \mathcal{C})$. Note also that we are following the usual convention that when no subscript is associated with the integral then the integration is assumed to be with respect to the entire set X.

We shall now develop some of the basic properties of the integral of functions $f \in M^+(X, \mathcal{C})$ similar to those obtained for simple functions $\psi \in M^+(X, \mathcal{C})$.

Theorem 2.2.20

Let f and g be elements of $M^+(X, \mathcal{C})$. If $f(x) \leq g(x)$ for each $x \in X$, then

$$\int f d\lambda \leq \int g d\lambda.$$

Proof: Note that by *Definition* (2.2.16),

$$\int f d\lambda = \sup \{\int \psi d\lambda : \psi \in M^+(X, \mathcal{C}) \text{ is simple and } 0 \leq \psi(x) \leq f(x) \text{ for } x \in X\} = \sup A_1$$

and

$$\int g d\lambda = \sup \{\int \psi d\lambda : \psi \in M^+(X, \mathcal{C}) \text{ is simple and } 0 \leq \psi(x) \leq g(x) \text{ for } x \in X\} = \sup A_2.$$

Since $f(x) \leq g(x)$ for each $x \varepsilon X$, $A_1 \subseteq A_2$ implying that $\sup A_1 \leq \sup A_2$ or that $\int f d\lambda \leq \int g d\lambda$.

Theorem 2.2.21

Let $f \varepsilon M^+(X, \mathcal{C})$ and let $E, F \varepsilon \mathcal{C}$. If $E \subseteq F$, then

$$\int_E f d\lambda \leq \int_F f d\lambda.$$

Proof: Note that by *Definition* (2.2.18)

$$\int_E f d\lambda = \int f \chi_E d\lambda \quad \text{and}$$

$$\int_F f d\lambda = \int f \chi_F d\lambda.$$

Since $E \subseteq F$, $(f\chi_E)(x) = (f\chi_F)(x) = f(x)$ for $x \varepsilon E$;

$(f\chi_E)(x) = 0 \leq (f\chi_F)(x)$ for $x \varepsilon F - E$;

$(f\chi_E)(x) = (f\chi_F)(x) = 0$ for $x \varepsilon X - F$.

Hence for $x \varepsilon X$ $(f\chi_E)(x) \leq (f\chi_F)(x)$. We may thus apply *Theorem* (2.2.20) to obtain that

$$\int f\chi_E d\lambda \leq \int f\chi_F d\lambda \quad \text{or}$$

$$\int_E f d\lambda \leq \int_F f d\lambda \quad \text{as}$$

was desired.

The next theorem of major interest in the development of the Lebesgue integral is the Monotone Convergence Theorem, a result attributable to B. Levi. We shall state this result without proof. The reader is referred to Bartle [2] for a detailed verification. This theorem provides the basis for many of the convergence properties of the Lebesgue integral and will also be of use in obtaining several important results in our later discussion. In order to understand the gist of the Monotone Convergence Theorem it will be necessary to review the meaning of several key terms from analysis.

Definition 2.2.22

Let (z_n) be a sequence of elements of $\overline{R}$. The *limit superior* of this sequence, denoted $\limsup z_n$, is defined by

$$\limsup z_n = \inf_m (\sup_{n \geq m} z_n).$$

The limit inferior of this sequence, denoted $\liminf z_n$, is defined by

$$\liminf z_n = \sup_m (\inf_{n \geq m} z_n).$$

If $\liminf z_n = \limsup z_n$ then their common value is defined to be the limit of the sequence (z_n), denoted $\lim z_n$.

Note 2.2.23

If (z_n) in a sequence of real numbers and $\lim z_n$ is finite, then *Note* (1.2.19) is equivalent to *Definition* (2.2.22).

For further clarification of the above definition the reader is referred to *Note* (2.2.15).

Definition 2.2.24

Let (f_n) be a sequence of functions in $M^+(X, \mathcal{C})$. (f_n) is said to be *monotone increasing* if $f_n(x) \le f_{n+1}(x)$ for each $x \in X$ and each natural number n.

Definition 2.2.25

Let (f_n) be a sequence of functions in $M^+(X, \mathcal{C})$ and let $f \in M^+(X, \mathcal{C})$. (f_n) is said to *converge* to f if for each $x \in X$, $\lim f_n(x) = f(x)$.

**Theorem* 2.2.26

(Monotone Convergence Theorem)

If (f_n) is a monotone increasing sequence of elements of $M^+(X, \mathcal{C})$ such that (f_n) converges to f in the sense of *Definition* (2.2.25), then $f \in M^+(X, \mathcal{C})$ and

$$\int f d\lambda = \lim \int f_n d\lambda.$$

Once again let us point out that since we are dealing with functions which map into $\overline{R}$ we are not requiring that either the integral on the left of this equation or the limit on the right be finite.

The following lemma allows one to "approximate" functions $f \in M^+(X, \mathcal{C})$ by an increasing sequence of simple functions in $M^+(X, \mathcal{C})$. This procedure is analogous to that used in Riemann theory when the integral of a non-negative function is approximated by a sequence of "step" functions.

Lemma 2.2.27

Let $f \in M^+(X, \mathcal{C})$. There exists a monotone increasing sequence (ψ_n) of simple functions in $M^+(X, \mathcal{C})$ converging to f.

Proof: Let n be an arbitrary but fixed natural number. For each integer k such that $0 \leq k \leq n2^n - 1$ define a set $E(k, n)$ by

$$E(k, n) = \{x \in X : \frac{k}{2^n} \leq f(x) < \frac{k+1}{2^n}\}.$$

For $k = n2^n$ define

$$E(k, n) = \{x \in X : f(x) \geq n\}.$$

Define ψ_n by

$$\psi_n(x) = \frac{k}{2^n} \qquad x \in E(k, n).$$

We shall show first that $\{E(k, n) : k = 0, 1, 2, \ldots, n2^n\}$ is a partition of X and that $\psi_n \in M^+(X, \mathcal{C})$. To this end let $x \in X$. Since $f \in M^+(X, \mathcal{C})$ either $f(x) \geq n$ or $0 \leq f(x) < n$. If $f(x) \geq n$, then $x \in E(n2^n, n) = f^{-1}([n, \infty))$. Note that

$\{E(k, n) : 0 \leq k \leq n2^n - 1\}$ is given by

$$E(0, n) = \{x : 0 \leq f(x) < \frac{1}{2^n}\} = f^{-1}([0, \infty)) \cap f^{-1}(-\infty, \frac{1}{2^n})$$

$$E(1, n) = \{x : \frac{1}{2^n} \leq f(x) < \frac{2}{2^n}\} = f^{-1}([\frac{1}{2^n}, \infty)) \cap f^{-1}(-\infty, \frac{2}{2^n})$$

$$E(2, n) = \{x : \frac{2}{2^n} \leq f(x) < \frac{3}{2^n}\} = f^{-1}([\frac{2}{2^n}, \infty)) \cap f^{-1}(-\infty, \frac{3}{2^n})$$

$$\vdots$$

$$E(n2^n - 2, n) = \{x : \frac{n2^n-2}{2^n} \leq f(x) < \frac{n2^n-1}{2^n}\} =$$

$$f^{-1}([\frac{n2^n-2}{2^n}, \infty)) \cap f^{-1}(-\infty, \frac{n2^n-1}{2^n})$$

$$E(n2^n-1, n) = \{x : \frac{n2^n-1}{2^n} \leq f(x) < \frac{n2^n}{2^n}\} =$$

$$f^{-1}([\frac{n2^n-1}{2^n}, \infty)) \cap f^{-1}(-\infty, n).$$

From this formulation it is obvious that $\{E(k, n) : k = 0, 1, 2, \ldots, n2^n\}$ is a partition of X and furthermore since f is measurable that $E(k, n) \in \mathcal{C}$ for each $k = 0, 1, 2, \ldots, n2^n$. Since we define ψ_n by

$$\psi_n(x) = \frac{k}{2^n} \quad \text{for} \quad x \in E(k, n)$$

ψ_n takes on only a finite number of non-negative real values and hence ψ_n is simple by *Definition* (2.2.1). The fact that ψ_n is measurable and is hence an element of $M^+(X, \mathcal{C})$ follows from the fact that for any real number α, $\psi_n^{-1}(\alpha, \infty)$ is either ϕ or else a finite union of sets $E(k, n)$. To verify that (ψ_n) is monotone increasing, let $x \in X$ and assume that k is such that $\psi_n(x) = \frac{k}{2^n}$ implying that $x \in E(k, n)$. Thus either $\frac{k}{2^n} \leq f(x) < \frac{k+1}{2^n}$ or else $\psi_n(x) = n$. If $\frac{k}{2^n} \leq f(x) < \frac{k+1}{2^n}$, then also

$$\frac{2k}{2^{n+1}} \leq f(x) < \frac{2(k+1)}{2^{n+1}} \quad \text{or} \quad \frac{2k}{2^{n+1}} \leq f(x) < \frac{2k+1+1}{2^{n+1}}.$$

By definition then, $x \in E(2k, n+1)$ or $x \in E(2k+1, n+1)$.

If $x \in E(2k, n+1)$ then $\psi_{n+1}(x) = \frac{2k}{2^{n+1}} = \frac{k}{2^n} = \psi_n(x)$.

If $x \in E(2k+1, n+1)$ then $\psi_{n+1}(x) = \frac{2k+1}{2^{n+1}}$. However, $2k+1 > 2k$ implying that $\frac{2k+1}{2^{n+1}} > \frac{2k}{2^{n+1}} = \frac{k}{2^n} = \psi_n(x)$. Hence in either case $\psi_n(x) \leq \psi_{n+1}(x)$. If $\psi_n(x) = n$, then $f(x) \geq n$. If also $f(x) \geq n+1$, then $\psi_{n+1}(x) = n+1 > \psi_n(x)$ as desired. If $n \leq f(x) < n+1$, then there exists a $k \in \{0, 1, 2, 3, \ldots, (n+1)2^{n+1} - 1\}$ such that $\frac{k}{2^{n+1}} \leq f(x) < \frac{k+1}{2^{n+1}}$. Note that $f(x) \geq n$ implies that $f(x) \geq \frac{n2^{n+1}}{2^{n+1}}$ and hence $k \in \{n2^{n+1}, n2^{n+1}+1, \ldots (n+1)2^{n+1} - 1\}$. That is, $k \geq n2^{n+1}$ implying that $\frac{k}{2^{n+1}} \geq n$. Hence $\psi_{n+1}(x) = \frac{k}{2^{n+1}} \geq n = \psi_n(x)$ and the proof that (ψ_n) is monotone increasing is complete. It remains only to show that $\lim \psi_n(x) = f(x)$ for each $x \in X$.

Choose $x \in X$ and consider $f(x)$. Either $f(x) = \infty$ or $f(x) = z < \infty$. If $f(x) = \infty$, then $f(x) > n$ for each natural number n. Hence by definition, $\psi_n(x) = n$ for each n and the sequence $(\psi_n(x)) = (1, 2, 3, 4, 5 \ldots)$. By *Definition* (2.2.22) $\lim \psi_n(x) = \infty = f(x)$. Assume now that $f(x) = z < \infty$. Choose $\varepsilon > 0$ and pick N_1 a natural number such that $\frac{1}{2^{N_1}} < \varepsilon$ and $z < N_1$. Note that since $\{E(k, N_1) : k = 0, 1, 2, \ldots, N_1 2^{N_1}\}$ is a partition of X there exists a k such that $x \in E(k, N_1)$ implying that

$$\frac{k}{2^{N_1}} \leq f(x) < \frac{k+1}{2^{N_1}}.$$

Hence $\psi_{N_1}(x) = \frac{k}{2^{N_1}}$ by definition and thus

$$|f(x) - \psi_{N_1}(x)| = f(x) - \frac{k}{2^{N_1}} < \frac{k+1}{2^{N_1}} - \frac{k}{2^{N_1}} = \frac{1}{2^{N_1}} < \varepsilon.$$

An analogous argument would yield that for any n such that $\frac{1}{2^n} < \varepsilon$, $|f(x) - \psi_n(x)| < \frac{1}{2^n}$. Since $n > N_1$ implies $\frac{1}{2^n} < \frac{1}{2^{N_1}} < \varepsilon$ we have that for $n > N_1$ $|f(x) - \psi_n(x)| < \frac{1}{2^n} < \frac{1}{2^{N_1}} < \varepsilon$ as was desired and the proof is complete.

We can now use the Monotone Convergence Theorem and *Lemma* (2.2.27) to derive some elementary properties of the integral of general functions $f \in M^+(X, \mathcal{C})$ analogous to those obtained in *Lemma* (2.2.10) and *Lemma* (2.2.11) for simple functions in $M^+(X, \mathcal{C})$.

Theorem 2.2.28

Let $f \in M^+(X, \mathcal{C})$ and $c \geq 0$. Then $cf \in M^+(X, \mathcal{C})$ and

$$\int cf d\lambda = c\int f d\lambda.$$

Proof: The fact that the theorem is true for $c = 0$ is trivial. Since $f \in M^+(X, \mathcal{C})$, there exists a monotone increasing sequence (ψ_n) of simple functions in $M^+(X, \mathcal{C})$ converging to f by *Lemma* (2.2.27). For $c > 0$, $(c\psi_n)$ converges to cf. By the Monotone Convergence Theorem $cf \in M^+(X, \mathcal{C})$ and

$$\int cf d\lambda = \lim \int c\psi_n d\lambda.$$

However, by *Lemma* (2.2.10) for each n, $\int c\psi_n d\lambda = c \int \psi_n d\lambda$. Thus

$$\int cf d\lambda = \lim c \int \psi_n d\lambda = c \lim \int \psi_n d\lambda.$$

Applying the Monotone Convergence Theorem to the limit on the right, we obtain that

$$\int cf d\lambda = c \int f d\lambda \text{ as was desired.}$$

Theorem 2.2.29

Let $f, g \in M^+(X, \mathcal{C})$. Then $f + g \in M^+(X, \mathcal{C})$ and

$$\int (f + g) d\lambda = \int f d\lambda + \int g d\lambda.$$

Proof: Since $f, g \in M^+(X, \mathcal{C})$ there exists a monotone increasing sequence (ψ_n) of simple functions from $M^+(X, \mathcal{C})$ converging to f and a monotone increasing sequence (ξ_n) of simple functions from $M^+(X, \mathcal{C})$ converging to g by *Lemma* (2.2.27). $(\psi_n + \xi_n)$ is a monotone increasing sequence of simple functions in $M^+(X, \mathcal{C})$ converging to $f + g$. By *Theorem* (2.2.26), $f + g \in M^+(X, \mathcal{C})$ and

$$\int (f + g) d\lambda = \lim \int (\psi_n + \xi_n) d\lambda .$$

By *Lemma* (2.2.11), for each n, $\int (\psi_n + \xi_n) d\lambda = \int \psi_n d\lambda + \int \xi_n d\lambda$. Hence $\int (f + g) d\lambda = \lim(\int \psi_n d\lambda + \int \xi_n d\lambda)$. Since for each n, $\int \psi_n d\lambda \geq 0$ and $\int \xi_n d\lambda \geq 0$, we can conclude that $\lim (\int \psi_n d\lambda + \int \xi_n d\lambda) = \lim \int \psi_n d\lambda + \lim \int \xi_n d\lambda$.

By applying *Theorem* (2.2.26) to the limits on the right, we obtain

$$\int (f + g)d\lambda = \int f d\lambda + \int g d\lambda$$

as was desired.

Thus far we have defined the integral only for functions $f \in M^+(X, \mathcal{C})$ and hence are severely limited in that we can speak of the integral only of non-negative functions measurable with respect to $(X, \mathcal{C}, \lambda)$. In order to define the integral of real valued functions f which are measurable with respect to $(X, \mathcal{C}, \lambda)$ it is necessary to consider the following definition:

Definition 2.2.30

Let $f : X \to \mathcal{R}$. The function f_+ mapping X into the non-negative real numbers defined by

$$f_+(x) = \sup \{f(x), 0\}$$

is called the *positive part of* f. The function f_- mapping X into the non-negative real numbers defined by

$$f_-(x) = \sup \{-f(x), 0\}$$

is called the *negative part of* f.

Theorem 2.2.31

Let $f : X \to \mathcal{R}$. Then

i) $f = f_+ - f_-$;

ii) $|f| = f_+ + f_-$;

iii) If f is measurable with respect to $(X, \mathcal{C}, \lambda)$ then f_+ and f_- are both measurable.

Proof: i) Let $x \in X$. Since f is a real valued function, either $f(x) = 0$, $f(x) < 0$ or $f(x) > 0$. If $f(x) = 0$, then $f_+(x) = f_-(x) = 0$ implying that $(f_+ - f_-)(x) = 0 = f(x)$ as was to be shown. If $f(x) < 0$, then $f_+(x) = 0$, and $f_-(x) = -f(x)$. Thus $(f_+ - f_-)(x) = 0 - (-f(x)) = f(x)$. If $f(x) > 0$, then $f_+(x) = f(x)$, and $f_-(x) = 0$. Thus again $(f_+ - f_-)(x) = f(x) - 0 = f(x)$.

ii) Note that if $f(x) \geq 0$, $|f(x)| = f(x) = f_+(x) - f_-(x)$ by i). However $f_-(x) = 0$ implying that we can write

$f(x) = f_+(x) + f_-(x)$ as desired. If $f(x) < 0$, then $|f(x)| = -f(x)$ and $f_-(x) = -f(x)$. Thus we have $|f(x)| = 0 - f(x) = f_+(x) + f_-(x)$.

iii) Assume that f is measurable and let $\alpha \in \mathcal{R}$. If $\alpha \leq 0$, then

$$f_+^{-1}(-\infty, \alpha) = f_-^{-1}(-\infty, \alpha) = \phi \in \mathcal{C}.$$

If $\alpha > 0$, then

$$f_+^{-1}(-\infty, \alpha) = f_+^{-1}[0, \alpha) = f^{-1}(-\infty, \alpha) \in \mathcal{C};$$

$$\begin{aligned} f_-^{-1}(-\infty, \alpha) &= f_-^{-1}[0, \alpha) \\ &= \{x : -f(x) \in [0, \alpha)\} \cup \{x : f(x) > 0\} \\ &= \{x : f(x) \in (-\alpha, 0]\} \cup \{x : f(x) > 0\} \\ &= f^{-1}(-\alpha, 0] \cup f^{-1}(0, \infty) \in \mathcal{C}. \end{aligned}$$

Note that thus far, the functions for which the Lebesgue integral has been defined have been <u>extended</u> real valued functions and hence we have allowed their integrals to take on the extended real value ∞. It will be convenient at this point to become more restrictive and consider only real valued functions and talk of Lebesgue integrals only for those functions whose "integral" is a finite real number. Note that for any real valued function f measurable with respect to $(X, \mathcal{C}, \lambda)$, the functions f_+ and f_- are in $M^+(X, \mathcal{C})$ and hence we may use *Definition* (2.2.16) to define the integrals of f_+ and f_-. We may use this fact to define what will be termed an integrable function.

Definition 2.2.32

Let f be a real valued function measurable with respect to $(X, \mathcal{C}, \lambda)$ such that $\int f_+ d\lambda < \infty$ and $\int f_- d\lambda < \infty$. f is said to be an *integrable function* and the integral of f with respect to λ, denoted $\int f d\lambda$, is defined by

$$\int f d\lambda = \int f_+ d\lambda - \int f_- d\lambda.$$

As was the case for non-negative measurable functions (see *Definition* (2.2.18)) it will occasionally be useful to speak of the integral of f over $E \in \mathcal{C}$, which we denote $\int_E f d\lambda$. By this we shall mean the following:

Definition 2.2.33

Let f be an integrable function in the sense of *Definition* (2.2.32) and let $E \in \mathcal{C}$. The integral of f over E, denoted $\int_E f d\lambda$, is defined by

$$\int_E f d\lambda = \int_E f_+ d\lambda - \int_E f_- d\lambda.$$

We shall not attempt at this point to develop the useful properties of the Lebesgue integral but shall refer the reader to Chapter Four, section three where many of these properties are detailed in a probabilistic setting. We shall however mention the following fact which is relevant to the above definition.

Theorem 2.2.34

If $f = f_1 - f_2$ where f_1 and f_2 are <u>any</u> non-negative functions such that $\int f_1 d\lambda < \infty$ and $\int f_2 d\lambda < \infty$ then $\int f d\lambda = \int f_1 d\lambda - \int f_2 d\lambda$.

Proof: Note that $f = f_1 - f_2 = f_+ - f_-$ by assumption and by *Theorem* (2.2.31). Hence $f_+ + f_2 = f_1 + f_-$. Since all of these functions are non-negative we may apply *Theorem* (2.2.29) to obtain

$$\int (f_+ + f_2) d\lambda = \int (f_1 + f_-) d\lambda \quad \text{or}$$

$$\int f_+ d\lambda + \int f_2 d\lambda = \int f_1 d\lambda + \int f_- d\lambda.$$

Rearranging terms we obtain

$$\int f_+ d\lambda - \int f_- d\lambda = \int f_1 d\lambda - \int f_2 d\lambda.$$

Thus $\int f d\lambda = \int f_1 d\lambda - \int f_2 d\lambda$.

The following set of exercises will be used in connection with the material of Chapter Four, section three to obtain some interesting results relative to the concept of expectation. They are also of interest in that they involve the application of many of the results obtained in this section. The first two results are based in part on

Fatou's Lemma, which comes as a result of the Monotone Convergence Theorem. We shall state this lemma without proof. For a detailed proof the reader is referred to Bartle [2].

2.2 *EXERCISES*

Definition:

Let (f_n) be a sequence of elements in $M^+(X, \mathcal{C})$. By $\liminf f_n$ we mean the function K defined by

$$K(x) = \liminf f_n(x)$$

where the limit inferior in the above equation is as defined in *Definition* (2.2.22).

**Fatou's Lemma*

Let (f_n) be a sequence of elements in $M^+(X, \mathcal{C})$ and let λ be a measure on $\mathcal{C}$. Then

$$\int (\liminf f_n) d\lambda \leq \liminf \int f_n d\lambda$$

1. Let $f \in M^+(X, \mathcal{C})$ and let λ be a measure on $\mathcal{C}$. If $E = \{x : f(x) > 0\}$ is an element of $\mathcal{C}$ and $\lambda(E) = 0$, then $\int f d\lambda = 0$.

Hint: Define $f_n(x) = n\,\chi_E(x)$. Show that this sequence satisfies the hypothesis of Fatou's Lemma and that $f \leq \liminf f_n$. Use Fatou's Lemma and *Theorem* (2.2.20) to complete the proof.

2. Let $f \in M^+(X, \mathcal{C})$ and let $F \in \mathcal{C}$. Let λ be a measure on $\mathcal{C}$ such that $\lambda(F) = 0$. Show that $\int_F f d\lambda = 0$.

Hint: Use *Exercise* (2.2.1) where $E = \{x : (f\chi_F)(x) > 0\}$.

3. Let $f \in M^+(X, \mathcal{C})$ and let λ be a measure on $\mathcal{C}$. Define a map γ on $\mathcal{C}$ by

$$\gamma(A) = \int_A f d\lambda.$$

Then γ is a measure on X.

Hint: Let (A_n) be a sequence of elements of $\mathcal{C}$ such that $A_i \cap A_j = \phi$, $i \neq j$. Define a sequence of functions f_n by $f_n = \sum_{i=1}^{n} f \chi_{A_i}$. Show that this sequence satisfies the conditions of *Theorem* (2.2.26) and that in fact $(f_n) \to f \chi_A$ where $A = \bigcup_{i=1}^{\infty} A_i$.

Note that the countably additive property of γ implies that

$$\gamma(A) \equiv \sum_{i=1}^{\infty} \gamma(A_i)$$

which in terms of integrals may be expressed as follows:

$$\int_A f d\lambda = \sum_{i=1}^{\infty} \int_{A_i} f d\lambda.$$

4. Prove that indicator function χ_A is non-measurable, where A is a non-measurable subset of X.

5. Let $(X, \mathcal{C}, \lambda)$ be a measure space and let $f_n : X \to \mathcal{R}$, $n = 1, 2, \ldots$ be a monotonic sequence of measurable functions such that $\lim f_n(x) = f(x)$. Prove $f(x)$ is measurable.

2.3 *THE RIEMANN-STIELTJES INTEGRAL*

We shall in this section indicate briefly how one goes about defining the Riemann-Stieltjes integral. The approach is similar to that used to define the Riemann integral by means of Riemann sums encountered in introductory calculus courses. Our interest here is simply in defining the integral as we shall not deal with this type of integration extensively in the material to follow. The reader is referred to Bartle [2] , and Widder [10] for a detailed discussion of this topic.

Definition 2.3.1

Let $J = [a, b]$ for a, b real. Let $x_0 = a < x_1 < x_2 \cdots < x_n = b$. Let $Q = \{[x_0, x_1], [x_1, x_2], \ldots, [x_{n-1}, x_n]\}$. Q is called a *partition* of J and is sometimes simply denoted by $Q = \{x_0, x_1, \ldots, x_n\}$.

Definition 2.3.2

Let h and g be bounded on some interval $J = [a, b]$ and let $Q = \{x_0, x_1, \ldots, x_n\}$ be a partition of J. A Riemann-Stieltjes sum of h with respect to g and corresponding to Q is the real number

$$S(Q; h; g) = \sum_{k=1}^{n} h(\varepsilon_k)[g(x_k) - g(x_{k-1})]$$ where

$\varepsilon_k \varepsilon [x_{k-1}, x_k]$ $k = 1, 2, \ldots n.$

Note 2.3.3

If $g(x) \equiv x$ then the above sum is simply the usual Riemann sum encountered in elementary calculus.

Definition 2.3.4

Let $Q = \{x_0, x_1, \ldots, x_n\}$ be a partition of $J = [a, b]$. The norm of Q, denoted $||Q||$, is defined by

$$||Q|| = \max_k \{x_k - x_{k-1}\}.$$

Definition 2.3.5

The Riemann-Stieltjes integral of h with respect to g from a to b denoted

$$\int_a^b hdg \quad \text{or} \quad \int_a^b h(x)dg(x)$$

is given by

$$\int_a^b h(x)dg(x) = \lim_{||Q||\to 0} \sum_{k=1}^{n} h(\varepsilon_k)[g(x_k)-g(x_{k-1})]$$

where $\varepsilon_k \varepsilon [x_{k-1}, x_k]$ $k = 1, 2, \ldots, n$ provided the limit exists.

Definition 2.3.6

$$\int_a^{\infty} hdg = \lim_{b\to\infty} \int_a^b hdg$$

and

$$\int_{-\infty}^{b} h\,dg = \lim_{a \to -\infty} \int_{a}^{b} h\,dg$$

provided these limits exist.

2.4 *MISCELLANEOUS TOPICS FROM ANALYSIS*

We shall present here a few miscellaneous results from mathematical analysis which will be useful later. Most of these results should be familiar to the reader and are summarized here for the sake of easy reference. They can be found in most basic texts on mathematical analysis or calculus.

**Theorem* 2.4.1

(Binomial Theorem)

Let r and s be any two real numbers and n a non-negative integer. Then

$$(r + s)^n = \sum_{k=0}^{n} \binom{n}{k} r^k s^{n-k} \qquad \text{where}$$

$$\binom{n}{k} = \frac{n!}{k!(n-k)!}, \quad 0! = 1, \quad n! = n(n-1)(n-2) \cdots (3)(2)(1)$$

for $n \geq 1$.

Definition 2.4.2

By

$$\int_{-\infty}^{\infty} f(x)\,dx \qquad \text{where}$$

the integral involved is a *Riemann integral* we mean

$$\lim_{b\to\infty} \int_0^b f(x)dx + \lim_{a\to-\infty} \int_a^0 f(x)dx$$

provided these limits exist. A similar definition holds for the Lebesgue-Stieltjes and Riemann-Stieltjes integrals which will be defined later.

Note 2.4.3

There are two distinct definitions commonly in use in defining $\int_{-\infty}^{\infty} f(x)dx$ in the case of a Riemann integral.

The first is called the "ordinary" value of the improper integral and is given above. The second, called the *Cauchy Principal Value* is given as follows:

$\int_{-\infty}^{\infty} f(x)dx = \lim_{h\to\infty} \int_{-h}^{h} f(x)dx$ provided this limit exists.

The two integrals agree in value whenever both exist but it is possible for the Cauchy Principal Value to exist when the ordinary value does not [3]. Whenever $f(x) > 0$ for all x or $f(x) < 0$ for all x the two will agree. Unless otherwise stated when we write $\int_{-\infty}^{\infty} f(x)dx$ we shall mean the ordinary value.

**Theorem* 2.4.4

The MacLaurin series expansion for e^z is given by

$$e^z = 1 + z + \frac{z^2}{2!} + \frac{z^3}{3!} + \cdots = \sum_{x=0}^{\infty} \frac{z^x}{x!}$$

for z real.

Definition 2.4.5

The function Γ defined on $(0, \infty)$ by

$$\Gamma(z) = \int_0^\infty x^{z-1} e^{-x}\, dx$$

is called the gamma function.

**Theorem* 2.4.6

Let n be a non-negative integer.

$$\Gamma(n + 1) = n!$$

**Theorem* 2.4.7

For $0 \le x \le 1$, $1 - x \le e^{-x}$.

**Theorem* 2.4.8

(*Cauchy Criterion For Series*)

If (a_n) is a sequence of real numbers, then $\sum_{n=1}^{\infty} a_n$ exists if and only if for each $\varepsilon > 0$ there exists a natural number N_1 such that

$$\left| \sum_{n=N_1}^{\infty} a_n \right| < \varepsilon.$$

**Theorem* 2.4.9

(*Abel's Partial Summation Formula*).

Consider the sequences (a_n) and (b_n). Let $A_n = \sum_{k=0}^{n} a_k$ for $n \ge 0$ and let $A_{-1} = 0$. Then if $0 \le p \le q$, we have

$$\sum_{n=p}^{q} a_n b_n = \sum_{n=p}^{q-1} A_n(b_n - b_{n+1}) + A_q b_q - A_{p-1} b_p.$$

Definition 2.4.10

Let $f : R \to R$. f is said to be *non-decreasing* if whenever $x_1 < x_2$, $f(x_1) \leq f(x_2)$.

Definition 2.4.11

Let $f : R \to R$. f is said to be even if $f(-x) = f(x)$ for each $x \in R$.

Definition 2.4.12

A sequence (a_n) of real numbers is said to be a *Cauchy sequence* if for every $\varepsilon > 0$ there exists a natural number N_1 such that for $n, m \geq N_1$, $|a_n - a_m| < \varepsilon$.

Theorem 2.4.13

(*Cauchy Criterion*)

Every Cauchy sequence in R converges.

2.5 *SUMMARY*

An introduction to the concepts of *elementary measure theory* was presented. The following terms were of major importance:

An ordered pair (X, C) such that X is a *set* and C a *σ-algebra* of subsets of X is called a *measurable space*. Let (X, C) be a measurable space. A *measure* is a set function defined on C such that

i) $\lambda(\phi) = 0$;

ii) $\lambda(A) \geq 0$ for each $A \in C$;

iii) λ is countably additive in the sense that if (A_n) is a sequence of elements of C

such that $A_i \cap A_j = \phi$ for $i \neq j$ then

$$\lambda(\bigcup_{i=1}^{\infty} A_n) = \sum_{i=1}^{\infty} \lambda(A_n).$$

An ordered triple $(X, \mathcal{C}, \lambda)$ where $(X, \mathcal{C})$ is a measurable space and λ is a measure on $\mathcal{C}$ is called a *measure space*.

Let $(X, \mathcal{C})$ be a measurable space and let f be a real valued function defined on X. f is said to be *measurable* if

$$f^{-1}(E) \in \mathcal{C}$$

for every Borel set E.

Emphasis was placed on the development of the *Lebesgue measure* on the real line and on the *measurable space* (R, β) where β is the collection of Borel sets. Equivalent ways to define the term measurable function were fully explored.

A complete development of the *Lebesgue integral* was presented with particular attention being paid to those properties of the integral which are pertinent to the study of *probability theory*. The *Riemann-Stieltjes* integral was defined and several other miscellaneous but useful results from real analysis were reviewed.

IMPORTANT TERMS IN CHAPTER TWO

measurable space
Borel subsets of the real line
extended real numbers
measure
measure space
finite measure space
σ-finite measure space
algebra
Borel subsets of (0, 1)
complete measure space
measurable function
simple function
indicator function
lim sup z_n
lim inf z_n
positive part of f
negative part of f
Lebesgue integral
Riemann-Stieltjes sum
norm of a partition
Riemann-Stieltjes integral
gamma function
non-decreasing function
even function
Cauchy sequence

REFERENCES
AND
SUGGESTIONS FOR FURTHER READINGS

[1] Bartle, R. G., The Elements of Real Analysis. New York: John Wiley and Sons, Inc., 1964.

[2] Bartle, R. G., The Elements of Integration. New York: John Wiley and Sons, Inc., 1966.

[3] Buck, R. C., Advanced Calculus. New York: McGraw Hill Book Company, Inc., 1956.

[4] Greever, John, Theory and Examples of Point Set Topology. Belmont, California: Wadsworth Publishing Company, Inc., 1967.

[5] Johnson, R. E. and Kiokemeister, F. L., Calculus with Analytic Geometry. Boston: Allyn and Bacon, Inc.

[6] Olmsted, J., Real Variables. New York: Appleton-Century-Crofts, 1959.

[7] Royden, H. L., Real Analysis. New York: The MacMillan Company, 1963.

[8] Rudin, Walter, Principles of Mathematical Analysis. New York: McGraw Hill Book Company, 1964.

[9] **Taylor, Angus E., General Theory of Functions and Integration. New York: Blaisdell Publishing Company, 1965.

[10] Widder, David V., Advanced Calculus. Englewood Cliffs, New Jersey: Prentice-Hall, Inc., 1961.

**This book is more advanced than the approach of the present text.

CHAPTER THREE

PROBABILITY AS AN AXIOMATIC SYSTEM

3.0 *INTRODUCTION*

We begin in this chapter our formal study of probability theory as an axiomatic system. Our first consideration shall be, as in the case when studying any mathematical topic, terminology and notation. The concepts and techniques of probability theory are closely related to ordinary set theory, mathematical analysis, and measure theory. However, since the study of probability theory in much of its earlier development was motivated by attempts to analyze games of chance, probabilistic concepts have in many instances names suggestive of this early beginning.

3.1 *PROBABILITY SPACES AND THEIR PROPERTIES*

We shall first define the term probability space and then present a dictionary of probabilistic terms used in this chapter together with their set or measure theoretic counterpart. Precise mathematical definitions can be found in Chapters One and Two.

Definition 3.1.1

A *probability space* $(\Omega, \mathcal{F}, P)$ is a *measure space of measure one*.

Throughout this chapter we shall assume that the *probability space* $(\Omega, \mathcal{F}, P)$ underlies our discussion.

The table which follows gives a precise comparison between *probabilistic* and *measure theoretic* concepts.

Note that by definition, Ω is a non-empty set, $\mathcal{F}$ is a σ algebra associated with Ω and P is a set function whose domain is $\mathcal{F}$ satisfying the following conditions:

i) $P[E] \geq 0$ for each $E \in \mathcal{F}$;

ii) $P[\Omega] = 1$;

iii) If $E_1, E_2, E_3, \ldots$ is a sequence of events such that $E_i \cap E_j = \phi$ for $i \neq j$ then

$$P\left[\bigcup_{i=1}^{\infty} E_i\right] = \sum_{i=1}^{\infty} P[E_i].$$

In measure theoretic terms property iii) is usually referred to as the *countably additive* property. This property is sometimes referred to as the *Extended Axiom of Addition* in probabilistic terminology. We shall refer to these properties as the *Axioms of Probability*. We shall also refer to any set function P defined on a measurable space $(\Omega, \mathcal{F})$ as a *probability measure* if $P[\Omega] = 1$.

Theorem 3.1.2

$P[\phi] = 0$.

Proof: Let $E_0 = \Omega$, $E_i = \phi$ for $i = 1, 2, 3, \ldots$.

Then

$$\Omega = \bigcup_{i=0}^{\infty} E_i \quad \text{and} \quad E_i \cap E_j = \phi \quad \text{if} \quad i \neq j.$$

DICTIONARY OF PROBABILISTIC TERMINOLOGY

	Set or Measure Theoretic Concepts		*Probabilistic Concept*	
	Notation	*Terminology*	*Notation*	*Terminology*
1.	$(\mathcal{U}, \mathcal{B}, \lambda)$	$(\mathcal{U}, \mathcal{B}, \lambda)$ is a *measure space* such that $\lambda[\mathcal{U}] = 1$	$(\Omega, \mathscr{F}, P)$	$(\Omega, \mathscr{F}, P)$ is a *probability space*
2.	$\mathcal{U}$	$\mathcal{U}$ is the *universal* set	Ω	Ω is the *sample space*
3.	$x \in \mathcal{U}$	x is an *element* of $\mathcal{U}$	$\omega \in \Omega$	ω is a *sample point* or an *outcome*
4.	$\mathcal{B}$	$\mathcal{B}$ is a σ *algebra*	$\mathscr{F}$	$\mathscr{F}$ is the collection of recognizable *events* associated with Ω
5.	$A \in \mathcal{B}$	A is an *element* of $\mathcal{B}$	$E \in \mathscr{F}$	E is an *event*
6.	ϕ	ϕ is the *empty* set	ϕ	ϕ is the *null* or *impossible* event
7.	$\mathcal{U} \in \mathcal{B}$	$\mathcal{U}$ is an *element* of $\mathcal{B}$	$\Omega \in \mathscr{F}$	Ω is the *sure* or *certain* event

DICTIONARY OF PROBABILISTIC TERMINOLOGY - Continued

	Set or Measure Theoretic Concepts		*Probabilistic Concept*	
	Notation	*Terminology*	*Notation*	*Terminology*
8.	$\{x\} \varepsilon \mathcal{B}$	$\{x\}$ is an *element* of $\mathcal{B}$	$\{\omega\} \varepsilon \mathcal{F}$	$\{\omega\}$ is an *elementary* or *simple* event
9.	$A, B \varepsilon \mathcal{B}$ and $A \cap B = \phi$	A and B are *disjoint*	E and $K \varepsilon \mathcal{F}$ and $E \cap K = \phi$	E and K are *mutually exclusive* events
10.	λ	λ is a *measure* on elements of $\mathcal{B}$	P	P is a *probability measure* on the collection of events of Ω
11.		λ is countably additive		P satisfies the Extended Axiom of Addition
12.		countable subadditivity		Boole's inequality

Thus $P[\Omega] = P[\bigcup_{i=0}^{\infty} E_i] = \sum_{i=0}^{\infty} P[E_i]$

$$= P[\Omega] + \sum_{i=1}^{\infty} P[\phi].$$

Thus $1 = 1 + \sum_{i=1}^{\infty} P[\phi]$ implying that

$$0 = \sum_{i=1}^{\infty} P[\phi] \quad \text{or} \quad P[\phi] = 0.$$

This theorem says essentially that if $E = \phi$, then $P[E] = 0$. The converse of this theorem is not necessarily true as shown by the following counterexample:

Example 3.1.3

Let $(\Omega, \mathcal{F}, P)$ be the probability space, such that $\Omega = (0, 1)$; $\mathcal{F}$ = collection of Borel subsets of Ω; P = Borel *measure*. Consider $P\{\omega:\omega = \frac{1}{n}, n = 2, 3, 4, 5, \ldots\}$. (See *Definition* (2.1.28) and *Theorem* (2.1.32)).

Theorem 3.1.4

Let $E_1, E_2, E_3, \ldots, E_n$ be elements of $\mathcal{F}$ such that $E_i \cap E_j = \phi$, $i \neq j$. Then $P[\bigcup_{i=1}^{n} E_i] = \sum_{i=1}^{n} P[E_i]$.

Proof: Consider the sequence $E_1, E_2, \ldots E_n, E_{n+1}, E_{n+2}, \ldots$ where $E_i = \phi$ $i > n$. Then $\bigcup_{i=1}^{n} E_i = \bigcup_{i=1}^{\infty} E_i$ and hence

$P[\bigcup_{i=1}^{n} E_i] = P[\bigcup_{i=1}^{\infty} E_i] = \sum_{i=1}^{\infty} P[E_i]$ by axiom iii). However

$$\sum_{i=1}^{\infty} P[E_i] = \sum_{i=1}^{n} P[E_i] + \sum_{i=n+1}^{\infty} P[E_i]$$

$$= \sum_{i=1}^{n} P[E_i] + \sum_{i=n+1}^{\infty} 0$$

by *Theorem* (3.1.2). Thus

$P[\bigcup_{i=1}^{n} E_i] = \sum_{i=1}^{n} P[E_i]$ as was to be shown.

Theorem 3.1.5

Let E_1, E_2 be elements of $\mathcal{F}$. If $E_1 \subseteq E_2$, then $P[E_1] \leq P[E_2]$.

Proof: Consider $E_2 = E_1 \cup (E_1' \cap E_2)$. Note that E_1' and $E_1' \cap E_2$ are events due to the fact that $\mathcal{F}$ is a σ algebra.

$P[E_2] = P[E_1] + P[E_1' \cap E_2]$ by *Theorem* (3.1.4). However $P[E_1' \cap E_2] \geq 0$ by axiom i). Hence $P[E_1' \cap E_2] + P[E_1] \geq P[E_1]$. By the transitivity property of the relation $\geq$, we obtain that $P[E_2] \geq P[E_1]$.

Theorem 3.1.6

For any event E, $P[E] \leq 1$.

Proof: Since E is an event $E \subseteq \Omega$. By *Theorem* (3.1.5) $P[E] \leq P[\Omega]$. By axion ii) $P[\Omega] = 1$. Hence $P[E] \leq 1$.

Note 3.1.7

This theorem implies that any probability measure is necessarily a σ-finite measure.

Theorem 3.1.8

Let E be any event. Then $P[E'] = 1 - P[E]$.

Proof: Note that $\Omega = E \cup E'$ and that $E \cap E' = \phi$ Thus by *Theorem* (3.1.4), $P[\Omega] = P[E] + P[E']$. However $P[\Omega] = 1$ by axiom ii) and we obtain $1 = P[E] + P[E']$ implying that $P[E'] = 1 - P[E]$.

Theorem 3.1.9

Let E_1 and E_2 be events such that $E_1 \subseteq E_2$. Then $P[E_2 - E_1] = P[E_2] - P[E_1]$.

Proof: Consider $E_2 = E_1 \cup (E_2 - E_1)$. Note that $E_1 \cap (E_2 - E_1) = \phi$. Hence $P[E_2] = P[E_1] + P[E_2 - E_1]$ implying that $P[E_2 - E_1] = P[E_2] - P[E_1]$.

Theorem 3.1.10

(Addition Principle)

Let E_1 and E_2 be events. Then

$$P[E_1 \cup E_2] = P[E_1] + P[E_2] - P[E_1 \cap E_2].$$

Proof: $E_1 \cup E_2 = (E_1 \cap E_2') \cup (E_1 \cap E_2) \cup (E_2 \cap E_1')$

$P[E_1 \cup E_2] = P[E_1 \cap E_2'] + P[E_1 \cap E_2] + P[E_2 \cap E_1']$ by *Theorem* (3.1.4). However

$$E_1' = (E_1 \cap E_2') \cup (E_1 \cap E_2) \quad \text{and}$$

$$E_2 = (E_2 \cap E_1') \cup (E_1 \cap E_2).$$

Thus by *Theorem* (3.1.4)

$$P[E_1] = P[E_1 \cap E_2'] + P[E_1 \cap E_2] \quad \text{and}$$

$$P[E_2] = P[E_2 \cap E_1'] + P[E_1 \cap E_2].$$

Solving for $P[E_1 \cap E_2']$ and $P[E_2 \cap E_1']$ respectively and substituting into the original equation we obtain $P[E_1 \cup E_2] = P[E_1] - P[E_1 \cap E_2] + P[E_1 \cap E_2] + P[E_2] - P[E_1 \cap E_2] = P[E_1] + P[E_2] - P[E_1 \cap E_2]$.

Theorem 3.1.11

i) If $E_1, E_2, E_3, \ldots$ is a sequence of events such that $E_1 \subseteq E_2 \subseteq E_3 \subseteq \ldots$, then

$$P[\bigcup_{n=1}^{\infty} E_n] = \lim_{n\to\infty} P[E_n].$$

ii) If $F_1, F_2, F_3, \ldots$ is a sequence of events such that $F_1 \supseteq F_2 \supseteq F_3, \ldots$, then

$$P[\bigcap_{n=1}^{\infty} F_n] = \lim_{n\to\infty} P[F_n].$$

We remark that a sequence of events satisfying the condition $E_1 \subseteq E_2 \subseteq E_3 \ldots$ will be referred to as an *increasing* sequence of events and a sequence satisfying $F_1 \supseteq F_2 \supseteq F_3 \supseteq \ldots$ will be referred to as a *decreasing* sequence of events.

Proof: i) Define $E_0 = \phi$ and consider the sequence of events

$$A_1 = E_1$$

$$A_2 = E_2 - E_1$$

$$\vdots$$

$$A_n = E_n - E_{n-1}$$

$$\vdots$$

Note that this sequence has the following properties:

$$A_i \cap A_j = \phi,\ i \neq j;\quad \bigcup_{n=1}^{\infty} A_n = \bigcup_{n=1}^{\infty} E_n;\quad \bigcup_{j=1}^{n} A_j = E_n.$$

Thus

$$P[\bigcup_{n=1}^{\infty} E_n] = P[\bigcup_{n=1}^{\infty} A_n] = \sum_{n=1}^{\infty} P[A_n]$$

by the *extended axiom of addition* or the *countable additivity* of P. By definition of infinite series we have

$$P[\bigcup_{n=1}^{\infty} E_n] = \lim_{m\to\infty} \sum_{n=1}^{m} P[A_n].$$

However, by *Theorem* (3.1.9)

$$P[A_n] = P[E_n] - P[E_{n-1}] \quad \text{for} \quad n \geq 1.$$

Thus,

$$\begin{aligned} P[\bigcup_{n=1}^{\infty} E_n] &= \lim_{m\to\infty} \sum_{n=1}^{m} [P(E_n) - P(E_{n-1})] \\ &= \lim_{m\to\infty} P[E_m] \\ &= \lim_{n\to\infty} P[E_n] \end{aligned}$$

as was to be shown.

ii) Let $(E_n) = (F_1 - F_n)$. Let $\omega \varepsilon E_n$. Then $\omega \varepsilon F_1$ but $\omega \notin F_n$. Since $F_{n+1} \subseteq F_n$, $\omega \notin F_{n+1}$. Thus by definition $\omega \varepsilon F_1 - F_{n+1} = E_{n+1}$. Hence $E_n \subseteq E_{n+1}$ and the sequence is increasing. By part i)

$$P[\bigcup_{n=1}^{\infty} E_n] = \lim_{n\to\infty} P[E_n] = \lim_{n\to\infty} P[F_1 - F_n].$$

However by *Theorem* (3.1.9) $P[F_1 - F_n] = P[F_1] - P[F_n]$.

Hence $P[\bigcup_{n=1}^{\infty} E_n] = \lim_{n\to\infty} (P[F_1] - P[F_n])$

$$= P[F_1] - \lim_{n\to\infty} P[F_n].$$

Now let $\omega \varepsilon \bigcup_{n=1}^{\infty} E_n$. Then $\omega \varepsilon E_n$ for some n implying that $\omega \varepsilon F_1$ but $\omega \notin F_n$. By definition of intersection $\omega \notin \bigcap_{n=1}^{\infty} F_n$. Thus $\omega \varepsilon F_1 - \bigcap_{n=1}^{\infty} F_n$. Similarly let $\omega \varepsilon F_1 - \bigcap_{n=1}^{\infty} F_n$. Then $\omega \varepsilon F_1$ but there exists an m such that $\omega \notin F_m$. Thus $\omega \varepsilon F_1 - F_m = E_m$. By definition of set union $\omega \varepsilon \bigcup_{n=1}^{\infty} E_n$. Hence $\bigcup_{n=1}^{\infty} E_n = F_1 - \bigcap_{n=1}^{\infty} F_n$ and $P[\bigcup_{n=1}^{\infty} E_n] = P[F_1 - \bigcap_{n=1}^{\infty} F_n]$. By *Theorem* (3.1.9) $P[\bigcup_{n=1}^{\infty} E_n] = P[F_1] - P[\bigcap_{n=1}^{\infty} F_n]$. Equating like terms we obtain $P[F_1] - \lim_{n\to\infty} P[F_n] = P[F_1] - P[\bigcap_{n=1}^{\infty} F_n]$ or $P[\bigcap_{n=1}^{\infty} F_n] = \lim_{n\to\infty} P[F_n]$.

The following corollary is an easy consequence of the preceeding theorem and is sometimes referred to as the *axiom of continuity*.

Corollary 3.1.12

If $F_1, F_2 \ldots$ is a decreasing sequence of events and if

$$\bigcap_{n=1}^{\infty} F_n = \phi,$$

then

$$\lim_{n\to\infty} P[F_n] = 0.$$

Example 3.1.13

Let $(\Omega, \mathcal{F}, P)$ be the probability space such that $\Omega = (0, 1)$; $\mathcal{F}$ = collection of Borel subsets of Ω; P = Borel measure. Define F_n by

$$F_n = (0, \frac{1}{n}).$$

By the axiom of continuity,

$$\lim_{n\to\infty} P[F_n] = 0.$$

See *Definition* (2.1.28).

Theorem 3.1.14

(Boole's Inequality)

Let $E_1, E_2, E_3, \ldots$ be a sequence of events. Then

$$P[\bigcup_{n=1}^{\infty} E_n] \leq \sum_{n=1}^{\infty} P[E_n].$$

Proof: Construct a sequence of events (A_n) as follows:

$$A_1 = E_1$$

$$A_2 = E_2 - E_1$$

$$A_3 = E_3 - (E_1 \cup E_2)$$

$$\vdots$$

$$A_n = E_n - (E_1 \cup E_2 \cdots \cup E_{n-1})$$

$$\vdots$$

Let $\omega \in \bigcup_{n=1}^{\infty} A_n$. Let $S = \{n : \omega \in A_n\}$. $S \neq \phi$ and by the Well Ordering Axiom of analysis there exists a least element N_1 in S. If $N_1 = 1$, we have $\omega \in E_1$ and hence $\omega \in \bigcup_{n=1}^{\infty} E_n$. If $N_1 \geq 2$ then $\omega \in E_{N_1} - (E_1 \cup E_2 \cup \ldots E_{N_1 - 1})$ in which case $\omega \in E_{N_1}$ and again we have $\omega \in \bigcup_{n=1}^{\infty} E_n$. Thus $\bigcup_{n=1}^{\infty} A_n \subseteq \bigcup_{n=1}^{\infty} E_n$. Now let $\omega \in \bigcup_{n=1}^{\infty} E_n$ and let $S_1 = \{n : \omega \in E_n\}$. $S_1 \neq \phi$ and by the Well Ordering Axiom there is a least element $N_2 \in S_1$. If $N_2 = 1$ then $\omega \in E_1 = A_1$ and hence $\omega \in \bigcup_{n=1}^{\infty} A_n$. If $N_2 \geq 2$ then $\omega \in E_{N_2}$ and $\omega \notin \bigcup_{n=1}^{N_2 - 1} E_n$ so that

$$\omega \in E_{N_2} - \bigcup_{n=1}^{N_2-1} = A_{N_2}.$$

Once again we obtain that $\omega \in \bigcup_{n=1}^{\infty} A_n$. Thus $\bigcup_{n=1}^{\infty} E_n \subseteq \bigcup_{n=1}^{\infty} A_n$ and we may conclude that $\bigcup_{n=1}^{\infty} E_n = \bigcup_{n=1}^{\infty} A_n$. To see that $A_i \cap A_j = \phi$ $i \neq j$ assume that there exist positive integers m and n and an element ω such that $\omega \in A_m \cap A_n$. Assume that $m > n$. By definition of A_m $\omega \in E_m$ but $\omega \notin E_n$. Hence by definition $\omega \notin A_n$ which is a contradiction. Thus $A_i \cap A_j = \phi$ $i \neq j$ as was to be shown. Now

$$P[\bigcup_{n=1}^{\infty} E_n] = P[\bigcup_{n=1}^{\infty} A_n] = \sum_{n=1}^{\infty} P[A_n].$$

However for each n $A_n \subseteq E_n$. By *Theorem* (3.1.5) $P[A_n] \leq P[E_n]$. Since we are dealing with non-negative series we may conclude that

$$\sum_{n=1}^{\infty} P[A_n] \leq \sum_{n=1}^{\infty} P[E_n].$$

Thus $P[\bigcup_{n=1}^{\infty} E_n] \leq \sum_{n=1}^{\infty} P[E_n]$ as was to be shown.

Note 3.1.15

The property obtained in *Boole's inequality* is sometimes of interest with respect to a general measure. In this context it is usually referred to simply as "*countable subadditivity*." The reader is referred to *Note* (2.1.24).

3.1 *EXERCISES*

1. Let $(\Omega, \mathcal{J})$ be a measurable space and let $P_1, P_2, P_3, \ldots, P_n$ be a collection of probability measures defined on $\mathcal{J}$. If $a_1, a_2, \ldots, a_n$ is a collection of non-negative real numbers such that $\sum_{i=1}^{n} a_i = 1$, then the function P^* defined on $\mathcal{J}$ by

$$P^*[E] = \sum_{i=1}^{n} a_i P_i[E]$$

is a probability measure on $\mathcal{J}$.

2. Let $(\Omega, \mathcal{J})$ be a measurable space and let $P_1, P_2, \ldots, P_n \ldots$ be a sequence of probability measures defined on $\mathcal{J}$. Prove that the function P^*

defined on $\mathcal{S}$ by

$$P^*[E] = \sum_{n=1}^{\infty} \frac{1}{2^n} P_n[E]$$

is a probability measure on $\mathcal{S}$.

3. Give an example to show that in a measure space $(X, \mathcal{C}, \lambda)$ with $\lambda(X) = \infty$ (for example Lebesgue measure on $\overline{R}$) that part ii) of *Theorem* (3.1.11) may not hold but that it does with the additional requirement that $\lambda(F_1) < \infty$.

4. Let $f : R \to R$ be a non-negative measurable function with $\int_{-\infty}^{\infty} f(x)dx = M$ where $0 < M < \infty$. Show that if for E a Borel set we define

$$P[E] = \frac{1}{M} \int_E f(x)dx$$

then P is a probability measure on (R, β).

3.2 *INDEPENDENCE AND CONDITIONAL PROBABILITY*

In this section we shall discuss the concept of independence of events. We shall first consider the concept of independence of two events E_1 and E_2 and shall then extend this idea to an arbitrary collection of events $\{E_\gamma : \gamma \in \Gamma\}$. In the latter case we shall distinguish two types of independence, pairwise independence and mutual independence. We shall also investigate the relationship between these concepts. We shall also introduce the concept of conditional probability and investigate its implications to the idea of independence.

Definition 3.2.1

Let E_1 and E_2 be events. We say that E_1 and E_2 are *independent events* if and only if

$$P[E_1 \cap E_2] = P[E_1]P[E_2].$$

Theorem 3.2.2

If E_1 and E_2 are independent and mutually exclusive events then either

$$P[E_1] = 0$$

or

$$P[E_2] = 0.$$

Proof: Assume that E_1 and E_2 are independent and mutually exclusive. Then $P[E_1 \cap E_2] = P[\phi] = 0$ by the definition of the term mutually exclusive events and *Theorem* (3.1.2). However $P[E_1 \cap E_2] = P[E_1]P[E_2]$ by *Definition* (3.2.1). Hence $P[E_1]P[E_2] = 0$ implying that either $P[E_1] = 0$ or $P[E_2] = 0$.

Theorem 3.2.3

If E_1 and E_2 are independent events, then E_1 and E_2' are also independent.

Proof: $E_1 \cap E_2' = E_1 - (E_1 \cap E_2)$. Hence $P[E_1 \cap E_2'] = P[E_1 - (E_1 \cap E_2)]$. However $E_1 \cap E_2 \subseteq E_1$ and by *Theorem* (3.1.9) $P[E_1 - (E_1 \cap E_2)] = P[E_1] - P[E_1 \cap E_2]$. Thus $P[E_1 \cap E_2'] = P[E_1] - P[E_1 \cap E_2]$. Since E_1 and E_2 are independent

$$\begin{aligned} P[E_1 \cap E_2'] &= P[E_1] - P[E_1]P[E_2] \\ &= P[E_1][1-P[E_2]] \\ &= P[E_1]P[E_2'] \end{aligned}$$

as was to be shown.

Theorem 3.2.4

If E_1 and E_2 are independent events, then so are E_1' and E_2'.

Proof: E_1 and E_2' are independent by *Theorem* (3.2.3). Applying *Theorem* (3.2.3) to these events we obtain that E_2' and E_1' are independent.

Note 3.2.5

Theorem (3.2.4) can be thought of as a corollary to *Theorem* (3.2.3). It can be shown to be a direct result of *Theorem* (3.2.3) by considering the fact that

$$E_1' \cap E_2' = E_1' - E_2 = E_1' - (E_1' \cap E_2).$$

Theorem 3.2.6

The null event and the certain event are independent of any event E.

Proof: $P[\phi \cap E] = P[\phi] = 0 = 0 \cdot P[E] = P[\phi]P[E]$. Furthermore $P[\Omega \cap E] = P[E] = 1 \cdot P[E] = P[\Omega]P[E]$.

We shall now extend the concept of independence to include more than two events. In fact, we shall consider independence relative to an arbitrary collection $\mathcal{C}$ of events.

Definition 3.2.7

Let $\mathcal{C} = \{E_\gamma : \gamma \varepsilon \Gamma\}$ be a collection of events. These events are said to be *mutually independent* if for every finite non-empty subset $\{E_1, E_2, . . . , E_n\}$

$$P[E_1 \cap E_2 \cap . . . \cap E_n] = P[E_1]P[E_2]. . . P[E_n].$$

Theorem 3.2.8

Let $\mathcal{C}$ be a collection of mutually independent events and let $\mathcal{B}_n$ be any finite subcollection of n elements of $\mathcal{C}$, $n \geq 2$. The number of equations which must be satisfied in order for $\mathcal{B}_n$ to be a collection of mutually independent events is given by

$$2^n - (n + 1).$$

Proof: In order to satisfy the definition of mutual independence we must show that the probability of the intersection of any subcollection of elements of $\mathcal{B}_n$ of cardinality two or more is equal to the product or their probabilities. There are $\binom{n}{2} + \binom{n}{3} + \binom{n}{4} + \cdot \cdot \cdot \cdot \binom{n}{n}$ such subcollections. However

$$\binom{n}{2} + \binom{n}{3} + \cdot \cdot \cdot \cdot \cdot \cdot \binom{n}{n} = \sum_{x=0}^{n} \binom{n}{x} - \binom{n}{0} - \binom{n}{1}$$

$$= 2^n - 1 - n$$
$$= 2^n - (n+1).$$

Theorem 3.2.9

Let $\mathcal{C}$ be a collection of mutually independent events. If each event in some subset $\mathcal{B}$ of $\mathcal{C}$ is replaced by its complement then the new collection $\mathcal{C}'$ so obtained is also a collection of mutually independent events.

Proof: We shall use induction on m for all n. That is, we shall show that for any collection of $m + n$ events

$$\{E_1, E_2, \ldots, E_n, F_1', F_2', \ldots, F_m'\}$$

where

$$\{E_1, E_2, \ldots, E_n\} \subseteq \mathcal{C} - \mathcal{B}$$

and

$$\{F_1, F_2, \ldots, F_m\} \subseteq \mathcal{B}$$

we have

$$P[E_1 \cap E_2 \cap \ldots \cap E_n \cap F_1' \cap F_2' \cap \ldots \cap F_m'] = \prod_{i=1}^{n} P[E_i] \prod_{j=1}^{m} P[F_j'].$$

To show that the theorem is true for $m = 1$ and all n, note that for any element F of $\mathcal{B}$ and elements $E_1, E_2, \ldots, E_n$ of $\mathcal{C} - \mathcal{B}$, F is independent of the event $E_1 \cap E_2 \cap \ldots \cap E_n$ due to the mutual independence of $\mathcal{C}$. Hence, F' is independent of the event

$E_1 \cap E_2 \cap \ldots \cap E_n$ by *Theorem* (3.2.3). Thus we can conclude that

$$P[E_1 \cap E_2 \cap \ldots \cap E_n \cap F'] = P[(E_1 \cap E_2 \cap \ldots \cap E_n) \cap F']$$
$$= P[E_1 \cap E_2 \cap \ldots \cap E_n]P[F']$$
$$= \prod_{i=1}^{n} P[E_i]P[F']$$

as was desired.

Now assume that the assertion is true for $m = k$ and *all* finite collections of elements of $\mathcal{C} - \mathcal{B}$. Note that

$$E_1 \cap E_2 \cap \ldots \cap E_n \cap F_1' \cap \ldots \cap F_k' \cap F_{k+1}' =$$
$$[E_1 \cap E_2 \cap \ldots \cap E_n \cap F_1' \cap \ldots \cap F_k']$$
$$- [E_1 \cap E_2 \cap \ldots \cap E_n \cap F_{k+1} \cap F_1' \cap F_2' \cap \ldots \cap F_k'],$$

and that

$$[E_1 \cap E_2 \cap \ldots \cap E_n \cap F_{k+1} \cap F_1' \cap F_2' \cap \ldots \cap F_k'] \subseteq$$
$$[E_1 \cap E_2 \cap \ldots \cap E_n \cap F_1' \cap \ldots \cap F_k'].$$

Thus by *Theorem* (3.1.9)

$$P[E_1 \cap E_2 \cap \ldots \cap E_n \cap F_1' \cap \ldots \cap F_k' \cap F_{k+1}'] =$$
$$P[E_1 \cap E_2 \cap \ldots \cap E_n \cap F_1' \cap \ldots \cap F_k']$$
$$- P[E_1 \cap E_2 \cap \ldots \cap E_n \cap F_{k+1} \cap F_1' \cap F_2' \cap \ldots \cap F_k'].$$

Using the inductive assumption that our assertion is true

for $m = k$ and *all* finite collections of elements of $\mathcal{C} - \mathcal{B}$, we have

$$P[E_1 \cap E_2 \cap \ldots \cap E_n \cap F_1' \cap \ldots \cap F_k' \cap F_{k+1}'] =$$

$$P[E_1]P[E_2] \ldots P[E_n]P[F_1'] \ldots P[F_k']$$

$$- P[E_1]P[E_2] \ldots P[E_n]P[F_{k+1}]P[F_1']P[F_2'] \ldots P[F_k']$$

$$= \left[P[E_1]P[E_2] \ldots P[E_n]P[F_1']P[F_2'] \ldots P[F_k']\right] \left[1 - P F_{k+1}]\right]$$

$$= P[E_1]P[E_2] \ldots P[E_n]P[F_1']P[F_2'] \ldots P[F_k']P[F_{k+1}'].$$

Thus the induction on m is complete and the theorem is proved.

Definition 3.2.10

Let $\mathcal{C} = \{E_\gamma : \gamma \in \Gamma\}$ be a collection of events. These events are said to be *pairwise independent* if

$$P[E_{\gamma_i} \cap E_{\gamma_j}] = P[E_{\gamma_i}]P[E_{\gamma_j}]$$

for each pair (γ_i, γ_j) where $\gamma_i, \gamma_j \in \Gamma$ and $\gamma_i \neq \gamma_j$.

Theorem 3.2.11

Mutual independence implies pairwise independence.

Example 3.2.12

This example will show that the converse of *Theorem* (3.2.11) is not necessarily true. That is, pairwise independence does not imply mutual independence. The example is due to S. N. Bernstein. Let

$$\Omega = \{\omega_1, \omega_2, \omega_3, \omega_4\}$$

where $P\{\omega_i\} = \frac{1}{4}$ for each i. Let

$$A = \{\omega_1, \omega_2\}, \quad B = \{\omega_1, \omega_3\}, \quad C = \{\omega_1, \omega_4\},$$

and consider the collection of events $\{A, B, C\}$. These events are pairwise independent but not mutually independent.

The following lemma is a well known lemma in probability theory. Its proof involves many of the concepts which have been discussed previously.

Lemma 3.2.13

(Borel Cantelli Lemma)

Let $E_1, E_2, E_3, \ldots$ be a sequence of events.

i) If $\sum_{n=1}^{\infty} P[E_n] < \infty$, then

$P\{\omega : \omega \in E_n$ for infinitely many natural numbers $n\} = 0$.

ii) If $\{E_1, E_2, \ldots .\}$ is a collection of mutually independent events and if

$\sum_{n=1}^{\infty} P[E_n] = \infty$, then

$P\{\omega : \omega \in E_n$ for infinitely many natural numbers $n\} = 1$.

Proof: Note that due to *Definition* (1.1.27)

$$\{\omega : \omega \in E_n \text{ for infinitely many natural numbers } n\} = \overline{\lim_{n\to\infty}} E_n.$$

Note also that due to *Theorem* (1.1.32),

$$\overline{\lim_{n\to\infty}} E_n = \bigcap_{n=1}^{\infty} \bigcup_{m=n}^{\infty} E_m.$$

Since

$$\bigcap_{n=1}^{\infty} \bigcup_{m=n}^{\infty} E_m \subseteq \bigcup_{m=n}^{\infty} E_m$$

for each n, we have that

$$P[\bigcap_{n=1}^{\infty} \bigcup_{m=n}^{\infty} E_m] \leq P[\bigcup_{m=n}^{\infty} E_m]$$

for each n by *Theorem* (3.1.5). However,

$$P[\bigcup_{m=n}^{\infty} E_m] \leq \sum_{m=n}^{\infty} P[E_m]$$

for each n by *Boole's Inequality*, *Theorem* (3.1.14). Hence,

$$P[\overline{\lim_{n\to\infty}} E_n] \leq \sum_{m=n}^{\infty} P[E_m]$$

for each n. Consider the sequence $(P[E_n])$ and the series

$$\sum_{i=1}^{\infty} P[E_i].$$

By hypothesis

$$\sum_{i=1}^{\infty} P[E_i] < \infty.$$

Hence, by the *Cauchy criterion* for series, *Theorem* (2.4.8), the sequence $(\sum_{m=n}^{\infty} P[E_m]$ converges to 0. Now, since for each n,

$$0 \leq P[\overline{\lim_{n\to\infty}} E_n] \leq \sum_{m=n}^{\infty} P[E_m]$$

the constant sequence, $(P[\overline{\lim}_{n\to\infty} E_n])$ must converge to 0. That is,

$$P[\overline{\lim}_{n\to\infty} E_n] = 0,$$

as was desired. To show that ii) is true note that $\{\omega : \omega \in E_n$ for infinitely many $n\} = \{\omega : \omega \in E_n$ for finitely many $n\}'$.

Hence, to show that

$$P[\overline{\lim}_{n\to\infty} E_n] = 1,$$

it is sufficient to show that

$$P\{\omega : \omega \in E_n \text{ for finitely many } n\} = 0.$$

However, $\{\omega : \omega \in E_n$ for finitely many $n\}$
$= \{\omega : \omega \in E_n'$ for all but a finite number of $E_n\}$
$= \overline{\lim}_{n\to\infty} E_n'$, by *Definition* (1.1.28).

Thus the problem is to show that

$$P[\underline{\lim}_{n\to\infty} E_n'] = 0.$$

Using *Theorem* (1.1.32) it is clear that

$$\underline{\lim}_{n\to\infty} E_n' = \bigcup_{n=1}^{\infty} \bigcap_{m=n}^{\infty} E_m' .$$

Hence,

$$P[\underline{\lim}_{n\to\infty} E_n'] = P[\bigcup_{n=1}^{\infty} \bigcap_{m=n}^{\infty} E_m']$$

$$\leq \sum_{n=1}^{\infty} P[\bigcap_{m=n}^{\infty} E_m'], \text{ by Boole's inequality.}$$

Note that for any positive integer k such that $k \geq n$

$$\bigcap_{m=n}^{\infty} E_m' \subseteq \bigcap_{m=n}^{k} E_m'.$$

Hence,

$$P[\bigcap_{m=n}^{\infty} E_m'] \leq P[\bigcap_{m=n}^{k} E_m'].$$

However,

$$P[\bigcap_{m=n}^{k} E_m'] = \prod_{m=n}^{k} P[E_m'], \text{ by } \textit{Theorem} \text{ (3.2.9) and our independence assumption}$$

$$= \prod_{m=n}^{k} [1 - P[E_m]], \text{ by } \textit{Theorem} \text{ (3.1.8)}$$

$$\leq \prod_{m=n}^{k} e^{-P[E_m]}$$

$$= e^{-\sum_{m=n}^{k} P[E_m]} \quad \text{by } \textit{Theorem} \text{ (2.4.7)}$$

Now, as $k \to \infty$,

$$\sum_{m=n}^{k} P[E_m] \to \infty$$

since

$$\sum_{n=1}^{\infty} P[E_n] = \infty.$$

Hence, for each n,

$$\lim_{k\to\infty} e^{-\sum_{m=n}^{k} P(E_m)} = 0 .$$

Thus,

$$P[\bigcap_{m=n}^{\infty} E_m'] \le 0$$

and

$$P[\underline{\lim}_{n\to\infty} E_n'] \le \sum_{n=1}^{\infty} P[\bigcap_{m=n}^{\infty} E_m'].$$

We must show that the series on the right goes to 0. Note that

$$\sum_{n=1}^{\infty} P[\bigcap_{m=n}^{\infty} E_m'] = P[\bigcap_{m=1}^{\infty} E_m'] + P[\bigcap_{m=2}^{\infty} E_m'] + P[\bigcap_{m=3}^{\infty} E_m'] + \ldots \quad .$$

The series on the right will go to zero if it can be shown that it is $\le \varepsilon$, for an arbitrarily small ε. This can be shown by arguing that

$$P[\bigcap_{m=1}^{\infty} E_m'] \le \varepsilon/2$$

$$P[\bigcap_{m=2}^{\infty} E_m'] \le \varepsilon/4$$

$$P[\bigcap_{m=3}^{\infty} E_m'] \le \varepsilon/8$$

$$\vdots$$

$$P[\bigcap_{m=n}^{\infty} E_m'] \leq \varepsilon/2^n,$$

$\cdot$
$\cdot$
$\cdot$

which implies that

$$\sum_{n=1}^{\infty} P[\bigcap_{m=n}^{\infty} E_m'] \leq \varepsilon/2 + \varepsilon/4 + \varepsilon/8 + \cdots$$

$$= \varepsilon\left(\frac{\frac{1}{2}}{1-\frac{1}{2}}\right) = \varepsilon.$$

Now to say that

$$\sum_{n=1}^{\infty} P[E_n] = \infty$$

implies that

i) $\sum_{n=m}^{\infty} P[E_n] = \infty,$ for any m;

and

ii) $\sum_{n=m}^{k} P[E_n]$, can be made arbitrarily large by picking k sufficiently large.

Now, for any n,

$$\bigcap_{m=n}^{\infty} E_m' \subseteq \bigcap_{m=n}^{k} E_m',$$

for any finite k. Hence,

$$P[\bigcap_{m=n}^{\infty} E_m'] \leq P[\bigcap_{m=n}^{k} E_m'],$$

for *any* finite k. However,

$$P[\bigcap_{m=n}^{k} E_m'] = \prod_{m=n}^{k} P[E_m'], \text{ by } \textit{Theorem } (3.2.9) \text{ and our independence criterion;}$$

$$= \prod_{m=n}^{k} [1 - P[E_m]], \text{ by } \textit{Theorem } (3.1.8)$$

$$\leq \prod_{m=n}^{k} e^{-P[E_m]}, \text{ by } \textit{Theorem } (2.4.6)$$

$$= e^{-\sum_{m=n}^{k} P[E_m]} .$$

Since

$$\sum_{n=1}^{\infty} P[E_n] = \infty$$

and since

$$\lim_{x\to\infty} e^{-x} = 0,$$

by remark ii) we can choose k large enough so that

$$e^{-\sum_{m=n}^{k} P[E_m]} < \varepsilon/2^n.$$

Thus, for all n, we can force $P[\bigcap_{m=n}^{\infty} E_m'] \leq \varepsilon/2^n$ as was desired.

Definition 3.2.14

Let E and F be events such that $P[F] \neq 0$. The *conditional probability* of E given F, denoted $P[E|F]$, is defined by

$$P[E|F] = \frac{P[E \cap F]}{P[F]} .$$

Theorem 3.2.15

Let F be an event such that $P[F] \neq 0$. The set function $P[\cdot|F]$ defined in *Definition* (3.2.14) is a probability measure.

Proof: Since P is a probability measure and $P[\cdot|F]$ is defined in terms of P, $P[E|F] \geq 0$ for each event E.

$$P[\Omega|F] = \frac{P[\Omega \cap F]}{P[F]} = \frac{P[F]}{P[F]} = 1.$$

Let $E_1, E_2, \ldots\ldots$ be a sequence of events such that $E_i \cap E_j = \phi$ for $i \neq j$.

Then $P[\bigcup_{i=1}^{\infty} E_i|F] = P[\bigcup_{i=1}^{\infty} E_i \cap F] / P[F]$

$$= P[\bigcup_{i=1}^{\infty} (E_i \cap F)] / P[F] .$$

Note that $(E_i \cap F) \cap (E_j \cap F) = \phi$ for $i \neq j$ and thus since P is a probability measure we may conclude that

$$P[\bigcup_{i=1}^{\infty} E_i|F] = \sum_{i=1}^{\infty} \frac{P[E_i \cap F]}{P[F]} = \sum_{i=1}^{\infty} P[E_i|F]$$

and the proof is complete.

The term "*two events are independent*" in the layman's sense of the word implies that the occurrence or non-occurrence of one event has no effect on the occurrence of the other. Likewise when we ask for the probability of some event E given F we are insinuating in the terminology that we have information that F has occurred and now wish to know the probability that E will occur based on this knowledge. If E and F are "independent" then it intuitively makes sense to assume that the probability of the occurrence of E is not affected by this knowledge. That this is indeed the case when we use the mathematical meaning of the terms is seen by the following theorem.

Theorem 3.2.16

Let events E and F be such that $P[F] > 0$. E and F are independent if and only if

$$P[E|F] = P[E],$$

Proof: Assume that E and F are independent. By *Definition* (3.2.14) and *Definition* (3.2.1)

$$P[E|F] = \frac{P[E \cap F]}{P[F]} = \frac{P[E]P[F]}{P[F]} = P[E].$$

To prove the converse, reverse the above argument.

Definition (3.2.14) can be rewritten to obtain the so called multiplication rule for two events. This rule can in turn be extended by mathematical induction to obtain a multiplication rule for any finite number of events. The latter is referred to as the *Law of Compound Probability.*

Theorem 3.2.17

(*Multiplication Rule*)

Let E and F be events such that $P[F] \neq 0$. Then $P[E \cap F] = P[E|F]P[F]$.

Theorem 3.2.18

(*Law of Compound Probability*)

Let $E_1, E_2, \ldots, E_n$ be events such that

$P[E_1 \cap E_2 \cap E_3 \ldots \cap E_n] > 0.$

Then

$$P[E_1 \cap E_2 \cap E_3 \ldots \cap E_n] = P[E_1]P[E_2|E_1]P[E_3|E_1 \cap E_2]$$
$$P[E_4|E_1 \cap E_2 \cap E_3] \ldots P[E_n|E_1 \cap E_2 \ldots E_{n-1}].$$

Proof: Note first that since $[E_1 \cap E_2 \cap \ldots E_n]$ is a subset of each of the sets

$E_1, E_1 \cap E_2, E_1 \cap E_2 \cap E_3, \ldots . E_1 \cap E_2 \cap \ldots E_{n-1}$

the condition $P[E_1 \cap E_2 \ldots \cap E_n] > 0$ together with *Theorem* (3.1.5) guarantees the existence of each of the probabilities on the right side of the equation. The truth of the theorem for $n = 2$ is guaranteed by *Theorem* (3.2.17). Let us assume that the result holds for any k events satisfying the given condition. Consider the $k+1$ events $E_1, E_2, \ldots . E_k, E_{k+1}$ and assume that

$P[E_1 \cap E_2 \ldots \cap E_k \cap E_{k+1}] > 0.$

$$P[E_1 \cap E_2 \ldots . \cap E_k \cap E_{k+1}] = P[(E_1 \cap E_2 \cap \ldots E_k) \cap E_{k+1}]$$
$$= P[E_1 \cap E_2 \ldots \cap E_k]P[E_{k+1}|E_1 \cap \ldots . E_k]$$

by *Theorem* (3.2.17). By the inductive assumption,

$$P[E_1 \cap E_2 \ldots . \cap E_k] =$$
$$P[E_1]P[E_2|E_1]P[E_3|E_1 \cap E_2] \ldots . P[E_k|E_1 \cap \ldots . E_{k-1}].$$

Hence $P[E_1 \cap E_2 \ldots . \cap E_{k+1}] =$

$$P[E_1]P[E_2|E_1]P[E_3|E_1 \cap E_2] \ldots .$$
$$P[E_k|E_1 \cap E_2 \ldots \cap E_{k-1}]P[E_{k+1}|E_1 \cap \ldots E_k]$$

and the induction is complete.

The following theorem is known as the *Law of Total Probability* and is instrumental in the development of the important formula of Bayes.

Theorem 3.2.19

(*Law of Total Probability*)

Let $E_1, E_2, \ldots E_n$ be a collection of events such that $P[E_i] > 0$ for each i, $P[\bigcup_{i=1}^{n} E_i] = 1$, and $E_i \cap E_j = \phi$ for $i \neq j$. Let F be an event. Then $P[F] = \sum_{i=1}^{n} P[F|E_i]P[E_i]$.

Proof: Note that since

$$P[\bigcup_{i=1}^{n} E_i] = 1, \quad P[(\bigcup_{i=1}^{n} E_i)'] = 0.$$

Express F as

$$F = [F \cap (\bigcup_{i=1}^{n} E_i)] \cup [F \cap (\bigcup_{i=1}^{n} E_i)'].$$

By *Theorem* (3.1.4)

$$P[F] = P[F \cap (\bigcup_{i=1}^{n} E_i)] + P[F \cap (\bigcup_{i=1}^{n} E_i)'].$$

Since

$$F \cap (\bigcup_{i=1}^{n} E_i)' \subseteq (\bigcup_{i=1}^{n} E_i)' \quad \text{by}$$

Theorem (3.1.5) we have

$$0 \leq P[F \cap (\bigcup_{i=1}^{n} E_i)'] \leq P[(\bigcup_{i=1}^{n} E_i)'] = 0.$$

Hence

$$P[F] = P[F \cap (\bigcup_{i=1}^{n} E_i)]$$

$$= P[\bigcup_{i=1}^{n} (E_i \cap F)].$$

Since $E_i \cap E_j = \phi$, $i \neq j$ we have also that

$(E_i \cap F) \cap (E_j \cap F) = \phi$ for $i \neq j$.

Thus by *Theorem* (3.1.4) $P[F] = \sum_{i=1}^{n} P[E_i \cap F]$.

By *Theorem* (3.2.17) $P[F] = \sum_{i=1}^{n} P[F|E_i]P[E_i]$.

Theorem 3.2.20

(*Bayes Theorem*)

Let $E_1, E_2, \ldots E_n$ be a collection of events such that $P[E_i] > 0$ for each i,

$P[\bigcup_{i=1}^{n} E_i] = 1$, and $E_i \cap E_j = \phi$ for $i \neq j$. Let F be an event such that $P[F] > 0$. Then

$$P[E_k|F] = \frac{P[F|E_k]P[E_k]}{\sum_{i=1}^{n} P[F|E_i]P[E_i]} \qquad k = 1, 2, \ldots n.$$

Proof: By *Definition* (3.2.14),

$$P[E_k|F] = \frac{P[E_k \cap F]}{P[F]}.$$

By *Theorem* (3.3.17) and *Theorem* (3.3.19)

$$P[E_k|F] = \frac{P[F|E_k]P[E_k]}{\sum_{i=1}^{n} P[F|E_i]P[E_i]} .$$

3.2 *EXERCISES*

1. Let $\{E_1, E_2, \ldots, E_n\}$ be a collection of mutually independent events.

   i) Show that the probability of the occurrence of at least one of the events is given by

   $$1 - \prod_{i=1}^{n} (1 - P[E_i]);$$

   ii) Show that the probability of the occurrence of exactly one of the events, say E_i, is given by

   $$P[E_i] \prod_{\substack{k=1 \\ k \neq i}}^{n} (1 - P[E_k]).$$

2. Let E_1 and E_2 be events such that $P[E_1] > 0$ and $P[E_2] > 0$. Show that if E_1 and E_2 are mutually exclusive then they are not independent but that the converse is not necessarily true.

3.3 *COMPLETENESS*

In this section we shall define what is meant by a *complete probability space*. This concept has been introduced in the general measure theoretic setting by *Theorem* (2.1.30). Its importance in the probabilistic context will

become evident in Chapter Five as probability spaces which
are not complete do allow for certain *"pathological"* and
rather unpleasant counterexamples to be constructed when
dealing with sequences of random variables. To avoid this
possibility the assumption of completeness is often a convenient one to make.

Definition 3.3.1

Let $(\Omega, \mathscr{f}, P)$ be a probability space. $(\Omega, \mathscr{f}, P)$ is
said to be *complete* if whenever $E \in \mathscr{f}$, $P[E] = 0$ and
$F \subseteq E$, then $F \in \mathscr{f}$ and $P[F] = 0$.

Note 3.3.2

To say that $(\Omega, \mathscr{f}, P)$ is a *complete probability space*
simply says that $\mathscr{f}$ contains every subset of any set of
probability measure zero.

Note 3.3.3

Given any probability space $(\Omega, \mathscr{f}, P)$ it is always
possible to expand the σ-algebra $\mathscr{f}$ to a σ-algebra $\mathscr{f}'$
in such a way that $\mathscr{f} \subseteq \mathscr{f}'$ and to extend the probability
measure P to a probability measure P' on $\mathscr{f}'$ in such a
way that $(\Omega, \mathscr{f}', P')$ is a complete probability space and
$P'[E] = P[E]$ for $E \in \mathscr{f}$. This can be seen from the following set of exercises.

3.3 *EXERCISES*

For each of the following exercises let $(\Omega, \mathcal{S}, P)$ be a probability space and let $\mathcal{Z} = \{E \in \mathcal{S} : P[E] = 0\}$.

1. Prove that $\mathcal{Z}$ is not a σ algebra but that if $E \in \mathcal{Z}$ and $F \in \mathcal{S}$ then $E \cap F \in \mathcal{Z}$ and that if (E_n) is a sequence of elements of $\mathcal{Z}$ then $\bigcup_{n=1}^{\infty} E_n \in \mathcal{Z}$.

2. Let $\mathcal{S}'$ be the collection of all subsets of Ω of the form

$$(E \cup Z_1) - Z_2$$

where $E \in \mathcal{S}$ and Z_1 and Z_2 are arbitrary elements of $\mathcal{Z}$. Prove that a set $F \in \mathcal{S}'$, if and only if it is an element of $\mathcal{S}$ and can be expressed in the form

$$F = E \cup K$$

where $E \in \mathcal{S}$ and $K \subseteq Z$ for some $Z \in \mathcal{Z}$.

Hint: The proof that if $F \in \mathcal{S}'$ then $F = E \cup K$ for $E \in \mathcal{S}$ and $K \subseteq Z$ for some $Z \in \mathcal{Z}$ is a straightforward application of *Theorem* (1.1.18) and the distribution property for set intersection. To show that if $F = E \cup K$ for $E \in \mathcal{S}$ and $K \subseteq Z$ for some $Z \in \mathcal{Z}$ then $F \in \mathcal{S}'$, consider the following diagram. Note that the shaded region is given by $Z - (E \cup K)$ and that

$$F = E \cup K = (E \cup Z) - [Z - (E \cup K)].$$

It remains only to argue that this representation satisfies the specified conditions.

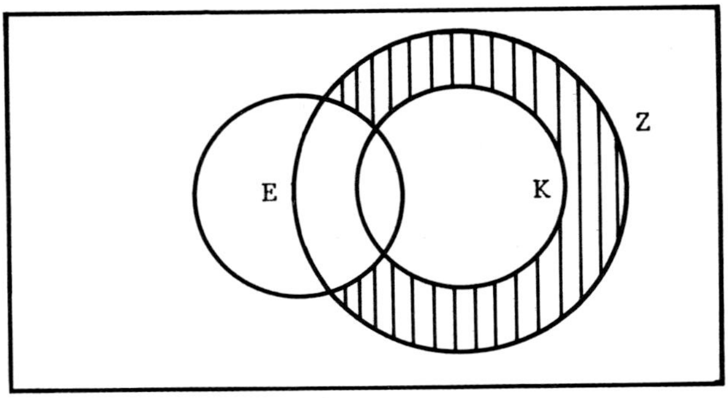

3. Show that the requirement that $F = E \cup K \in \mathcal{J}$ is a necessary condition in exercise two.

4. Show that $\mathcal{J}'$ defined in *Exercise* (3.3.2) is a σ algebra of subsets of Ω.

5. Define a map P' on $\mathcal{J}'$ of *Exercise* (3.3.2) by

$$P'[E \cup K] = P[E]$$

for $E \in \mathcal{J}$ and $K \subseteq Z$ for $Z \in \mathcal{Z}$. Show that P' is well defined and is a probability measure on $\mathcal{J}'$. *Hint*: To show that P' is well defined assume that $F \in \mathcal{J}'$ can be expressed as

$$F = E_1 \cup K_1 = E_2 \cup K_2$$

where $E_1, E_2 \in \mathcal{J}$ and $K_1 \subseteq Z_1, K_2 \subseteq Z_2$ for Z_1 and Z_2 elements of $\mathcal{Z}$. To show that P' is well defined, it is necessary to show that

$$P[E_1] = P'[E_1 \cup K_1] = P'[E_2 \cup K_2] = P[E_2].$$

Consider the fact that

$$E_1 \subseteq E_1 \cup K_1 = E_2 \cup K_2 \subseteq E_2 \cup Z_2$$

and

$$E_2 \subseteq E_2 \cup K_2 = E_1 \cup K_1 \subseteq E_1 \cup Z_1$$

and apply *Theorems* (3.1.5) and (3.1.14) to each of the above. The verification that P' is a probability measure on $\mathcal{F}'$ is straightforward.

6. Show that the probability space $(\Omega, \mathcal{F}', P')$ constructed in *Exercises* (3.3.1) to (3.3.5) is complete.

3.4 *SUMMARY*

The study of *probability* as an *axiomatic system* was begun by defining a *probability space* to be a measure space $(\Omega, \mathcal{F}, P)$ of measure one. Elements of $\mathcal{F}$ have been termed "*events*" and P has been referred to as a "*probability*" measure. The following "*Axioms of Probability*" were obtained from the measure theoretic properties of P:

Axiom i) $P[E] \geq 0$ for each $E \in \mathcal{F}$;

Axiom ii) $P[\Omega] = 1$;

Axiom iii) If $E_1, E_2, \ldots$ is a sequence of events such that $E_i \cap E_j = \phi$ for $i \neq j$ then

$$P\left[\bigcup_{i=1}^{\infty} E_i\right] = \sum_{i=1}^{\infty} P[E_i].$$

From these axioms many theorems were obtained concerning the behavior of P. Notable among these were the *Axioms of Continuity* and *Boole's Inequality*.

The concept of *independence* among events was introduced and two types of independence were considered, namely *mutual independence* and *pairwise independence*. More precisely, let $\mathcal{C} = \{E_\gamma : \gamma \in \Gamma\}$ be a collection of events. These events are said to be *mutually independent* if for every finite non-empty subset $\{E_1, E_2, \ldots, E_n\}$

$$P[E_1 \cap E_2 \ldots \cap E_n] = P[E_1]P[E_2] \ldots P[E_n].$$

These events are said to be *pairwise independent* if

$$P[E_{\gamma_i} \cap E_{\gamma_j}] = P[E_{\gamma_i}]P[E_{\gamma_j}]$$

for each pair (γ_i, γ_j) where $\gamma_i, \gamma_j \in \Gamma$ and $\gamma_i \neq \gamma_j$. Several theorems relative to *independence* were presented among them the well known *Borel Cantelli Lemma*. The idea of *conditional probability* was introduced and its relationship to independence was explored.

The chapter was concluded by a brief discussion of *complete probability spaces*.

IMPORTANT TERMS IN CHAPTER THREE

probability space
Axioms of probability
probability measure
mutually exclusive events
sample space
sample point
outcome
null event
sure event
elementary or simple event
decreasing sequence of events
increasing sequence of events
mutually independent
pairwise independent
conditional probability
complete probability space

REFERENCES AND SUGGESTIONS FOR FURTHER READINGS

[1] **Burrill, C. W., Measure, Integration and Probability. New York: McGraw-Hill Book Company, 1972.

[2] Gnedenko, B. V., The Theory of Probability. New York: Chelsea Publishing Company, 1966.

[3] Tsokos, C. P., Probability Distributions: An Introduction to Probability Theory with Applications. Belmont, California: Wadsworth Publishing Company, Inc., 1972.

[4] **Tucker, H. G., A Graduate Course in Probability. New York: Academic Press, 1967.

**These books are more advanced than the approach of the present text.

CHAPTER FOUR

ONE DIMENSIONAL RANDOM VARIABLES

4.0 INTRODUCTION

In this chapter we shall introduce the concept of a
random variable defined on a probability space $(\Omega, \mathfrak{f}, P)$.
We shall stress in particular what will be termed *"one
dimensional"* random variables. As the name implies it is
possible to extend the ideas presented here in an obvious
manner to form what are termed *n-dimensional random variables*.
We shall defer this discussion until Chapter Six.

Among the topics considered in this chapter are:
*functional forms of random variables; distribution and
density functions; continuous and discrete random variables;
calculating probabilities; and expectation.*

4.1 ONE DIMENSIONAL RANDOM VARIABLES

The following table is provided to draw a parallel
between measurable functions and Lebesgue integration and
their probabilistic counterparts. Once this connection has
been made many of the results of Chapter Two can simply be
rephrased in a probabilistic sense to obtain results basic to
probability theory.

DICTIONARY OF PROBABILISTIC TERMINOLOGY

Measure Theoretic Concepts: Notation	Measure Theoretic Concepts: Terminology	Probabilistic Concepts: Notation	Probabilistic Concepts: Terminology
1. X	X is a measurable function with respect to the measure space $(\Omega, \mathcal{F}, P)$	X	X is a random variable with respect to the probability space $(\Omega, \mathcal{F}, P)$
2. $\int X \, d P < \infty$	X is Lebesgue integrable over Ω with respect to P	$E[X] < \infty$	X has finite expectation
3. S holds P. a. e.	$(\Omega, \mathcal{F}, P)$ is a measure space and proposition S holds for almost all $\omega \varepsilon \Omega$	S holds P. a. e.	$(\Omega, \mathcal{F}, P)$ is a probability space and proposition S holds with probability one.
4. $\int \lvert X \rvert^2 \, d P < \infty$	the measurable function X is an element of the Lebesgue space $L_2(\Omega, \mathcal{F}, P)$	$E[\lvert X \rvert^2] < \infty$	the random variable X has finite second moment

Definition 4.1.1

Let $(\Omega, \mathcal{S}, P)$ be a probability space. Let $X : \Omega \to R$. X is said to be a *one dimensional random variable* if and only if X is measurable.

Note 4.1.2

There is a brief discussion of the term *measurable* in Chapters One and Two. The reader is referred to *Definition* (1.2.12), *Theorem* (1.2.13), *Definition* (2.1.33) to *Theorem* (2.1.38) in order to see some of the mathematical implications of this definition. *Theorem* (2.1.38) is of particular interest in this respect as it will allow the term "*random variable*" to be defined in several different but equivalent forms.

Throughout this discussion we shall be assuming the existence of an underlying probability space $(\Omega, \mathcal{S}, P)$. When we say that X_1 and X_2 are random variables we shall be assuming that they are each random variables associated with the same probability space $(\Omega, \mathcal{S}, P)$ although this fact may not always be stated explicitly.

Theorem 4.1.3

Let X be a constant function defined on Ω. X is a random variable.

Proof: Assume that $X(\omega) \equiv c$ for some arbitrary but fixed real number c. Let $B \subseteq R$ be any Borel set. Either $c \in B$ or $c \notin B$. If $c \in B$, then $X^{-1}(B) = \Omega$. If $c \notin B$ then $X^{-1}(B) = \phi$. Since $\Omega \in \mathcal{S}$ and $\phi \in \mathcal{S}$, X is measurable and hence is a random variable.

Theorem 4.1.4

Let $c \in R$ and assume $c > 0$. Let X be a random variable. $c\,X$ is a random variable.

Proof: Let α be an arbitrary but fixed real number. We must show that $(cX)^{-1}(-\infty, \alpha) \in \mathcal{J}$. However $(cX)^{-1}(-\infty, \alpha)$

$$= \{\omega : (cX)(\omega) < \alpha\}$$
$$= \{\omega : X(\omega) < \alpha/c\}$$
$$= X^{-1}(-\infty, \alpha/c).$$

Since X is a random variable, by *Theorem* (2.1.38) $X^{-1}(-\infty, \alpha/c) \in \mathcal{J}$ as was to be shown.

Theorem 4.1.5

If X is a random variable then $-X$ is a random variable.

Proof: Let α be an arbitrary but fixed real number. We must show that $(-X)^{-1}(-\infty, \alpha) \in \mathcal{J}$. However,

$$(-X)^{-1}(-\infty, \alpha) = \{\omega : (-X)(\omega) < \alpha\}$$
$$= \{\omega : X(\omega) > -\alpha\}$$
$$= X^{-1}(-\alpha, \infty).$$

Since X is a random variable we have by *Theorem* (2.1.38) that $X^{-1}(-\alpha, \infty) \in \mathcal{J}$ as was desired.

Corollary 4.1.6

If X is a random variable then $c\,X$ is a random variable for any real number c.

Proof: We have this result for $c = 0$ by *Theorem* (4.1.3) and for $c > 0$ by *Theorem* (4.1.4). If $c < 0$ then $c = -(-c)$ where $-c > 0$. Write $cX = -(-c)X$. By *Theorem* (4.1.4) $(-c)X$ is a random variable and by *Theorem* (4.1.5) $-(-c)X$ is a random variable.

Theorem 4.1.7

Let X_1 and X_2 be random variables. $X_1 - X_2$ is a random variable.

Proof: We shall show that for each real number α,

$$\{\omega : (X_1 - X_2)(\omega) < \alpha\} \in \mathcal{S}$$

which by *Theorem* (2.1.38) is sufficient to show that $X_1 - X_2$ is a random variable. Let

$$T = \{\omega : (X_1 - X_2)(\omega) < \alpha\} = \{\omega : X_1(\omega) < X_2(\omega) + \alpha\}.$$

For each $\omega \in T$ there exists a rational number r_ω such that

$$X_1(\omega) < r_\omega < X_2(\omega) + \alpha.$$

(Note that there are actually an infinite number of such rational numbers. We need only one corresponding to each ω). Let

$$S = \{r_\omega : X_1(\omega) < r_\omega < X_2(\omega) + \alpha\}.$$

Let

$$G = \bigcup_{r_\omega \in S} [X_1^{-1}(-\infty, r_\omega) \cap X_2^{-1}(r_\omega - \alpha, \infty)].$$

Since X_1 and X_2 are random variables, for each

$$r_\omega \in S,\ X_1^{-1}(-\infty, r_\omega) \text{ and } X_2^{-1}(r_\omega - \alpha, \infty)$$

are elements of $\mathcal{S}$ by *Theorem* (2.1.38). Since $\mathcal{S}$ is a σ algebra, for each $r_\omega \in S$, $X_1^{-1}(-\infty, r_\omega) \cap X_2^{-1}(r_\omega - \alpha, \infty)$ is an element of $\mathcal{S}$. Note that $S \subseteq Q$ where Q is the set of rational numbers. Since Q is countable, S is also countable by *Theorem* (1.1.13). Thus G is the countable union of elements of $\mathcal{S}$ and hence is itself an element of $\mathcal{S}$. We shall now show that $T = G$ which will complete the proof. Let $z \in T$. Then there exists an $r_z \in S$ such that

$$X_1(z) < r_z < X_2(z) + \alpha.$$

Thus, $z \in X_1^{-1}(-\infty, r_z)$ and $z \in X_2^{-1}(r_z - \alpha, \infty)$. This in turn implies that

$$z \in X_1^{-1}(-\infty, r_z) \cap X_2^{-1}(r_z - \alpha, \infty)$$

and hence also $z \in G$. We thus obtain that $T \subseteq G$. Now let $z \in G$. Then there exists an $r_z \in S$ such that

$$z \in X_1^{-1}(-\infty, r_z) \cap X_2^{-1}(r_z - \alpha, \infty).$$

This implies that $X_1(z) < r_z$ and $X_2(z) > r_z - \alpha$. Hence we obtain

$$X_1(z) < r_z < X_2(z) + \alpha$$

which implies that $z \in T$ and that $G \subseteq T$. By *Definition* (1.1.6) $G = T$ and the proof is complete.

Corollary 4.1.8

Let X_1 and X_2 be random variables. $X_1 + X_2$ is a random variable.

Proof: Note that $X_1 + X_2 = X_1 - (-X_2)$. By *Theorem* (4.1.5), $-X_2$ is a random variable. By *Theorem* (4.1.7), $X_1 - (-X_2)$ is a random variable.

Theorem 4.1.9

Let $X_1, X_2, X_3, \ldots$ be random variables. $X_1 + X_2 + \cdots + X_n$ is a random variable for every natural number n.

Proof: The method of proof is induction. The theorem is true for $n = 1$ by assumption and for $n = 2$ by *Corollary* (4.1.8). Assume that the theorem is true for $n = k$. That is, assume that $X_1 + X_2 + \cdots X_k$ is a random variable. Write $X_1 + X_2 + \cdots X_k + X_{k+1} = (X_1 + X_2 + \cdots X_k) + X_{k+1}$ and apply *Corollary* (4.1.8).

Corollary 4.1.10

Let $c_1, c_2, \ldots, c_n$ be real numbers and $X_1, X_2, \ldots, X_n$ be random variables. $c_1X_1 + c_2X_2 + \cdots + c_nX_n$ is a random variable.

Proof: For each $i = 1, 2, 3, \ldots n$, c_iX_i is a random variable by *Corollary* (4.1.6).

$\sum_{i=1}^{n} c_iX_i$ is a random variable by *Theorem* (4.1.9).

Note 4.1.11

Corollary (4.1.10) can be simply expressed by saying

that any linear combination of random variables is also a random variable.

Theorem 4.1.12

If X is a random variable, then X^2 is a random variable.

Proof: Let $\alpha \in R$. If $\alpha \leq 0$ then $\{\omega : X^2(\omega) < \alpha\} = \phi$. Note that $\phi \in \mathcal{F}$. If $\alpha > 0$, then $\{\omega : X^2(\omega) < \alpha\}$

$$= \{\omega : |X(\omega)| < \sqrt{\alpha}\}$$
$$= \{\omega : -\sqrt{\alpha} < X(\omega) < \sqrt{\alpha}\}$$
$$= X^{-1}(-\infty, \sqrt{\alpha}) \cap X^{-1}(-\sqrt{\alpha}, \infty).$$

By *Theorem* (2.1.38) $X^{-1}(-\infty, \sqrt{\alpha})$ and $X^{-1}(-\sqrt{\alpha}, \infty) \in \mathcal{F}$ and hence also $X^{-1}(-\infty, \sqrt{\alpha}) \cap X^{-1}(-\sqrt{\alpha}, \infty) \in \mathcal{F}$.

Theorem 4.1.13

If X_1 and X_2 are random variables then X_1X_2 is a random variable.

Proof: $X_1X_2 = \frac{1}{4}[(X_1 + X_2)^2 - (X_1 - X_2)^2]$. By *Corollary* (4.1.8), $X_1 + X_2$ is a random variable and by *Theorem* (4.1.7), $X_1 - X_2$ is a random variable. By *Theorem* (4.1.12), $(X_1 + X_2)^2$ and $(X_1 - X_2)^2$ are random variables. By *Theorem* (4.1.7) $(X_1 + X_2)^2 - (X_1 - X_2)^2$ is a random variable. By *Corollary* (4.1.6), $\frac{1}{4}[(X_1 + X_2)^2 - (X_1 - X_2)^2]$ is a random variable as was desired.

Theorem 4.1.14

If X is a random variable, then X^n is a random variable for each natural number n.

Proof: The method of proof is induction. By *Theorem* (4.1.12) the theorem is true for $n = 2$. Assume that the theorem is true for $n = k$. That is, assume that X^k is a random variable. Write $X^{k+1} = X^k \cdot X$ and apply *Theorem* (4.1.12).

Note 4.1.15

Corollary (4.1.6), *Theorem* (4.1.14) and *Theorem* (4.1.9) imply that if X is a random variable then any polynomial in X is also a random variable.

Theorem 4.1.16

Let X be a random variable and let $f : R \to R$ be Borel measurable. The composite function $f \circ X$ is a random variable. (We shall sometimes write $f(X)$ to indicate a composite map reserving juxtaposition to indicate simple products.)

Proof: Let B be a Borel set.

$$\begin{aligned}(f \circ X)^{-1}(B) &= \{\omega : f(X(\omega)) \in B\} \\ &= \{\omega : X(\omega) \in f^{-1}(B)\} \\ &= X^{-1}(f^{-1}(B)).\end{aligned}$$

Since f is Borel measurable, $f^{-1}(B)$ is a Borel set by *Note* (2.1.40). Thus the fact that X is a random variable and hence measurable implies that $X^{-1}(f^{-1}(B)) \in \mathcal{F}$ by *Definition* (2.1.34).

Corollary 4.1.17

Let X be a random variable and let $f : R \to R$ be continuous. The composite function $f \circ X$ is a random variable.

Proof: By *Theorem* (2.1.42) f continuous from $R \to R$ implies that f is Borel measurable. Hence by *Theorem* (4.1.16) f o X is a random variable.

Note 4.1.18

Since any polynomial $f : R \to R$ is continuous, *Theorem* (4.1.17) implies that *Theorem* (4.1.3), *Theorem* (4.1.4) *Theorem* (4.1.5), *Corollary* (4.1.6), *Theorem* (4.1.12) and *Theorem* (4.1.14) can be thought of as special cases of *Corollary* (4.1.17). These theorems were proved independently of *Corollary* (4.1.17) to give the reader some experience in working with the definition of the term random variable.

Theorem 4.1.19

Let X be a random variable and let $f : R \to R$ be monotonic. The composite function f o X is a random variable.

Proof: By *Theorem* (2.1.43) f monotonic implies that f is Borel measurable. Hence by *Theorem* (4.1.16) f o X is a random variable.

4.1 *EXERCISES*

1. Let $X_1, X_2, \ldots, X_n$ be random variables.
   i) Show that for each i, $|X_i|$ is a random variable;
   ii) Show that for each i, $[X_i]$ is a random variable where [ ] denotes the usual "greatest integer" function of elementary calculus;
   iii) Show that $Z = \dfrac{X_1 + X_2 + \cdots + X_n}{n}$ is a random variable for any natural number n;

iv) Show that for any constant c and any natural number $n > 1$, $\sum_{i=1}^{n} \frac{(X_i - c)^2}{n - 1}$ is a random variable;

v) Show that for each i, e^{X_i} is a random variable.

2. Let f be a real valued function with domain $A \subseteq R$, A a Borel set. Let $X : \Omega \to A$. If f is continuous then $f \circ X$ is a random variable.

Hint: The reader is referred to *Note* (1.2.17). Consider $(-\infty, \alpha)$ for α real and show that $(f \circ X)^{-1}(-\infty, \alpha) = X^{-1}(f^{-1}(-\infty, \alpha)) \in \mathcal{S}$ where $(\Omega, \mathcal{S}, P)$ is the underlying probability space.

3. Use *Exercise* (4.1.2) to verify that

i) If $X : \Omega \to R_+$ where R_+ denotes the positive real numbers, then $\ell n\, X$ is a random variable;

ii) If $X : \Omega \to R_+ \cup \{0\}$, then $\sqrt{X}$ is a random variable;

iii) If $X : \Omega \to R - \{0\}$, then $\frac{1}{X}$ is a random variable.

4.2 *DISTRIBUTION FUNCTIONS: CONTINUOUS AND DISCRETE RANDOM VARIABLES*

In this section we shall define what is known as the *cumulative distribution function* or simply the *distribution function* F of a one dimensional random variable X and examine some of its general properties. As in section one, we shall be assuming that $X_1, X_2, \ldots$ represent one dimensional random variables with respect to some underlying probability space $(\Omega, \mathcal{F}, P)$ unless specified otherwise. We shall also consider the two most common types of random variables namely *discrete random variables* and *continuous random variables*. We shall spend some time in investigating some computational aspects of these two types since they are the types which arise most often in *physical problems*.

Definition 4.2.1

Let $(\Omega, \mathcal{F}, P)$ be a probability space and let X be a random variable defined on Ω. Let $F : \mathcal{R} \to \mathcal{R}$ be defined by

$$F(x) = P\{\omega : X(\omega) \leq x\} = P[X^{-1}(-\infty, x]].$$

F is called the *cumulative distribution function* or *distribution function* of the random variable X.

Note 4.2.2

Since P maps into the closed unit interval $[0, 1]$, the range of F will also be a subset of $[0, 1]$. Notationally we shall use the following convention when dealing with F:

$$F(x) = P\{\omega : X(\omega) \leq x\} = P[X \leq x].$$

We shall also write

$$P[X = x], \; P[X \geq x], \; P[X \neq x],$$

for

$$P\{\omega : X(\omega) = x\}, \; P\{\omega : X(\omega) \geq x\}, \; P\{\omega : X(\omega) \neq x\},$$

respectively.

Example 4.2.3

Let $(\Omega, \mathcal{J}, P)$ be a probability space and let $X : \Omega \to \{-1, 1\}$ such that

$$P[X = -1] = P[X = 1] = \frac{1}{2} .$$

Then F is given by

$$F(x) = \begin{cases} 0 & \text{if } x < -1 \\ \frac{1}{2} & \text{if } -1 \leq x < 1 \\ 1 & \text{if } 1 \leq x . \end{cases}$$

The graph of F is a step function, whose graph is given below.

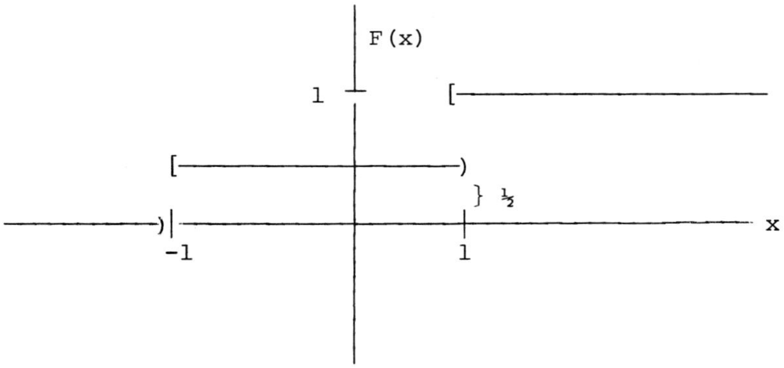

Example 4.2.4

Let $\Omega = \{H, T\}$. Let $\mathcal{J} = \rho_\Omega$. Define P on $\mathcal{J}$ by

$$P\{H\} = P\{T\} = \frac{1}{2}.$$

Define random variables X_1 and X_2 on Ω by

$$X_1(H) = X_2(T) = 1$$

and

$$X_1(T) = X_2(H) = 0 \ .$$

Let us indicate the distribution functions for X_1 and X_2 by F_{X_1} and F_{X_2} respectively. Then

$$F_{X_1}(x) = F_{X_2}(x) = \begin{cases} 0 & \text{if } x < 0 \\ \frac{1}{2} & \text{if } 0 \leq x < 1 \\ 1 & \text{if } 1 \leq x \ . \end{cases}$$

Note 4.2.5

The random variables X_1 and X_2 of the preceeding example are not identical, however their distribution functions are identical. Hence it is evident that distribution functions do <u>not</u> characterize random variables.

The following theorem summarizes the principle properties of the distribution function F. The reader is referred to *Definition* (1.2.21) to *Theorem* (1.2.23) for a discussion of right hand continuity.

Theorem 4.2.6

Let $(\Omega, \mathcal{F}, P)$ be a probability space, X a random variable defined on Ω, F the distribution function of X. F satisfies the following properties:

i) If $x_1 \leq x_2$, then $F(x_1) \leq F(x_2)$; (that is, F is a monotone increasing function)

ii) F is continuous from the right at each point $x_o \varepsilon R$;

iii) $\lim_{x\to\infty} F(x) = 1$ and $\lim_{x\to-\infty} F(x) = 0$.

Proof: i) Assume that $x_1 \leq x_2$. Then

$$(-\infty, x_1] \subseteq (-\infty, x_2]$$

and hence

$$X^{-1}(-\infty, x_1] \subseteq X^{-1}(-\infty, x_2].$$

By *Theorem* (3.1.5)

$$P[X^{-1}(-\infty, x_1]] \leq P[X^{-1}(-\infty, x_2]]$$

and thus by *Definition* (4.2.1)

$$F(x_1) \leq F(x_2).$$

ii) Let $x_o \varepsilon R$. Let (x_n) be sequence of real numbers such that $x_1 > x_2 > x_3 \cdots > x_n \cdots > x_o$, and $(x_n) \to x_o$. Let $F_n = X^{-1}(-\infty, x_n] - X^{-1}(-\infty, x_o]$. It is obvious that

$$F_1 \supseteq F_2 \supseteq F_3 \supseteq \cdots$$

and that $\bigcap_{n=1}^{\infty} F_n = \phi$. Hence by *Corollary* (3.1.12),

$$\lim_{n\to\infty} P[F_n] = 0.$$

Note that since

$$(-\infty, x_o] \subseteq (-\infty, x_n],$$

$$X^{-1}(-\infty, x_o] \subseteq X^{-1}(-\infty, x_n]$$

and we have by *Theorem* (3.1.9) that for each n

$$P\ X^{-1}(-\infty, x_n] - X^{-1}(-\infty, x_o]] = P[F_n]$$

$$= P[X^{-1}(-\infty, x_n]] - P[X^{-1}(-\infty, x_o]].$$

This in turn implies that for each n,

$$P[F_n] = F(x_n) - F(x_o).$$

Letting $n \to \infty$ we obtain that

$$\lim_{n\to\infty} P[F_n] = \lim_{n\to\infty} F(x_n) - \lim_{n\to\infty} F(x_o)$$

or

$$0 = \lim_{n\to\infty} F(x_n) - F(x_o).$$

Hence

$$(F(x_n)) \to F(x_o)$$

and by *Theorem* (1.2.23) F is continuous from the right at x_o.

There are two types of random variables commonly encountered in a beginning study of probability, namely random variables of the discrete type and those of the continuous type. We shall define each of these here and consider some specific examples of each.

Definition 4.2.7

Let $(\Omega, \mathcal{f}, P)$ be a probability space and let X be a random variable defined on Ω. X is said to be discrete if its range is countable.

Note 4.2.8

The random variables of *Examples* (4.2.3) and (4.2.4) are discrete.

Definition 4.2.9

Let X be a discrete random variable. The non-negative function f defined on $\mathcal{R}$ by

$$f(x) = P[X = x]$$

is called the density function or the probability density function of X.

Discrete random variables are commonly defined by simply stating the defining equation for its density function. We shall mention here several of the most common discrete densities.

Examples 4.2.10

i) A random variable X is said to have a point binomial distribution with parameter p if its density f is given by

$$f(x) = \begin{cases} p^x(1-p)^{1-x} & , \quad x = 0,1 \\ 0 & , \quad \text{elsewhere} \end{cases}$$

where $0 < p < 1$. A parameter is simply some constant associated with the distribution which when known completely specifies the density.

ii) A random variable X is said to have a binomial distribution with parameters n and p if its density f is given by

$$f(x) = \begin{cases} \binom{n}{x} p^x(1-p)^{n-x} & , \; x = 0, 1, 2, \ldots n \\ 0 & , \; \text{elsewhere} \end{cases}$$

where $0 < p < 1$ and n is a positive integer.

iii) A random variable X is said to have a Poisson distribution with parameter s if its density f is given by

$$f(x) = \begin{cases} \dfrac{e^{-s} s^x}{x!} & , \; x = 0, 1, 2, 3, \ldots \\ 0 & , \; \text{elsewhere} \end{cases}$$

where $s > 0$.

Definition 4.2.11

Let $(\Omega, \mathcal{F}, P)$ be a probability space and let X be a random variable defined on Ω. X is said to be of the continuous type if there exists a non-negative function f on $\mathcal{R}$, called the density function of X, such that

$$F(x) = P[X \leq x] = \int_{-\infty}^{x} f(t)dt$$

where the integral involved is the ordinary Riemann integral.

Note 4.2.12

By applying the fundamental theorem of integral calculus, it is evident that $F' = f$ whenever f is continuous. Since f is Riemann integrable, f is continuous a.e. (Lebesgue measure) so that $f' = f$ a.e. Thus, f is not determined everywhere but a.e.

Examples 4.2.13

i) A random variable X is said to have a *uniform distribution* over $[a, b]$ if its density f is given by

$$f(x) = \begin{cases} \frac{1}{b - a} & , \quad a < x < b \\ 0 & , \quad \text{elsewhere}; \end{cases}$$

ii) A random variable X is said to have a *Gaussian normal* or simply a *normal distribution* with parameters s and a if its density f is given by

$$f(x) = \frac{1}{s\sqrt{2\pi}} e^{-\frac{1}{2}\left(\frac{x - a}{s}\right)^2}, \quad x \in R, \; a \in R, \; s > 0;$$

iii) A random variable X is said to have a *Cauchy distribution* if its density f is given by

$$f(x) = \frac{1}{\pi(1 + x^2)} \qquad , \quad x \in \mathcal{R};$$

iv) A random variable X is said to have an *exponential distribution* with parameter α if its density f is given by

$$f(x) = \begin{cases} \frac{1}{\alpha} e^{-\frac{x}{\alpha}} & , \quad x > 0,\ \alpha > 0 \\ 0 & , \quad \text{elsewhere.} \end{cases}$$

In many instances it is desirable to consider statements of the form

$$P[X^{-1}(B)] = P[X \in B]$$

for some Borel set B of the real line and some random variable X. In the discrete case this can be a simple problem as can be seen from the following theorem if the density f for X is readily available.

Theorem 4.2.14

Let X be a random variable of the discrete type with density function f and range $\{x_\gamma : \gamma \in \Gamma\}$. Let B be a Borel set. Then

$$P[X \in B] = \sum_{x_\gamma \in B} f(x_\gamma).$$

Proof: Let $\{x_\gamma : \gamma \in \Gamma\}$ be the range of X. Let $E_\gamma = \{\omega : X(\omega) = x_\gamma\}$. Let $\Delta \subseteq \Gamma$ be such that $\delta \in \Delta$ if and only if $x_\delta \in B$.

Express $\{\omega : X(\omega) \varepsilon B\}$ as

$\bigcup_{\delta\varepsilon\Delta} E_\delta$. Then
$$P[X \varepsilon B] = P[\bigcup_{\delta\varepsilon\Delta} E_\delta]$$
$$= \sum_{\delta\varepsilon\Delta} P[E_\delta]$$
$$= \sum_{\delta\varepsilon\Delta} f(x_\delta)$$
$$= \sum_{x_\gamma\varepsilon B} f(x_\gamma).$$

Note 4.2.15

If X is a discrete random variable with range $\{x_\gamma : \gamma \varepsilon \Gamma\}$, density f and distribution function F, then

$$F(x) = \sum_{x_\gamma \leq x} f(x_\gamma).$$

Examples 4.2.16

i) Let X be a discrete random variable with a point *binomial density* as given in *Example* (4.2.10). Then F is given by

$$F(x) = 0, \quad x < 0$$
$$F(x) = 1-p, \quad 0 \leq x < 1$$
$$F(x) = 1, \quad 1 \leq x;$$

The graph of F is a step function.

ii) Let X be a discrete random variable with a *binomial density* as given in *Example* (4.2.10). Let $n = 10$, $p = \frac{1}{3}$. Then

$$P[X \in \{2, 3, 4\}] = \sum_{x \in \{2,3,4\}} \binom{10}{x} \left(\frac{1}{3}\right)^x \left(\frac{2}{3}\right)^{10-x} .$$

iii) Let X be a discrete random variable with a *Poisson distribution* as given in *Example* (4.2.10).

$$P[X \in R] = \sum_{x=0}^{\infty} f(x) = 1.$$

iv) For any discrete random variable X with density f and range $\{x_\gamma : \gamma \in \Gamma\}$,

$$\sum_{\gamma \in \Gamma} f(x_\gamma) = 1.$$

Verification

i), ii), and iii) are obvious. To see iv) let

$E_\gamma = \{\omega : X(\omega) = x_\gamma\}$.

$\{E_\gamma : \gamma \in \Gamma\}$ is a partition of Ω and thus

$$\begin{aligned} 1 - P[\Omega] = P[\bigcup_{\gamma \in \Gamma} E_\gamma] &= \sum_{\gamma \in \Gamma} P[E_\gamma] \\ &= \sum_{\gamma \in \Gamma} P[X = x_\gamma] \\ &= \sum_{\gamma \in \Gamma} f(x_\gamma). \end{aligned}$$

If the density function f for the discrete random variable X is not readily available or if X is a random variable of the continuous type, then the problem of expressing $P[X \in B]$ is more complex. It can be handled theoretically with the aid of the Lebesgue and Lebesgue-Stieltjes integrals as we will demonstrate. In dealing with random variables of the continuous type the type of

Borel set of most interest is one of the form $[a, b]$, a and b real. We shall therefore consider this case in more detail.

Recall that the event $X \in B$ is equivalent to the event $\{\omega : X(\omega) \in B\}$ and hence that the statement $P[X \in B]$ is a statement framed with respect to the abstract probability space $(\Omega, \mathcal{F}, P)$. Thus it is possible to calculate $P[X \in B]$ by means of the Lebesgue integral as follows:

Theorem 4.2.17

$$P[X \in B] = \int_{X^{-1}(B)} d\,P.$$

Proof: $P[X \in B] = P[X^{-1}(B)]$ by notational convention. By *Definition* (2.2.18)

$$\int_{X^{-1}(B)} d\,P = \int \chi_{X^{-1}(B)}\, d\,P.$$

By *Definition* (2.2.7)

$$\int \chi_{X^{-1}(B)}\, d\,P = 1 \cdot P[X^{-1}(B)] + 0 \cdot P[\Omega - X^{-1}(B)]$$

$$= P[X^{-1}(B)].$$

It will be helpful to construct a probability space $(\mathcal{R}, \beta, P')$ on the real line corresponding to $(\Omega, \mathcal{F}, P)$ in the sense that we can use knowledge of the value of $P'[B]$ to determine the value of $P[X \in B]$. This step will allow us to calculate $P[X \in B]$ by use of the Lebesgue-Stieltjes integral. The space $(\mathcal{R}, \beta, P')$ is called the induced probability space and its construction is described in the following theorem. Note that the theorem is applicable to any type of random variable.

Theorem 4.2.18

Let $(\Omega, \mathcal{J}, P)$ be a probability space and let X be a random variable defined over Ω. Define a function P' over the set β of Borel sets of $\mathcal{R}$ by

$$P'[B] = P[X^{-1}(B)].$$

The ordered triple $(\mathcal{R}, \beta, P')$ is a probability space.

Proof: We need only verify that P' satisfies the three axioms of probability. To this end let $B \varepsilon \beta$. $P[B] = P[X^{-1}(B)]$. Since X is a random variable $X^{-1}(B) \varepsilon \mathcal{J}$ implying that $P[X^{-1}(B)]$ is defined and is non-negative. Note that $P'[\mathcal{R}] = P[X^{-1}(\mathcal{R})] = P[\Omega] = 1$. Let $B_1, B_2, \ldots\ldots$ be a sequence of Borel sets such that $B_i \cap B_j = \phi$ for $i \neq j$.

$$\begin{aligned} P'[\bigcup_{i=1}^{\infty} B_i] = P[X^{-1}(\bigcup_{i=1}^{\infty} B_i)] &= P[\bigcup_{i=1}^{\infty} X^{-1}(B_i)] \\ &= \sum_{i=1}^{\infty} P[X^{-1}(B_i)] \\ &= \sum_{i=1}^{\infty} P'[B_i]. \end{aligned}$$

The space $(\mathcal{R}, \beta, P')$ is simpler to work with than the original probability space in that we are now dealing with the real line rather than with an abstract set Ω. It is still not at all clear as to how one would actually go about determining numerically the value of $P'[B]$ and hence also the value of $P[X \varepsilon B]$. In order to see how this can be accomplished we shall need to consider an important result from real analysis. This result is stated here without proof [1].

*Theorem** 4.2.19

If $g : \mathcal{R} \to \mathcal{R}$ is a monotone increasing function which is continuous from the right then there exists a unique measure μ_g defined on the collection of Borel sets of $\mathcal{R}$ such that for all $a, b \in \mathcal{R}$

$$\mu_g((a, b]) = g(b) - g(a)$$

$$\mu_g((-\infty, b]) = g(b) - \lim_{x \to -\infty} g(x)$$

$$\mu_g((a, \infty)) = \lim_{x \to \infty} g(x) - g(a)$$

$$\mu_g((-\infty, \infty)) = \lim_{x \to \infty} g(x) - \lim_{x \to -\infty} g(x).$$

Note 4.2.20

Observe that since g is monotone increasing each of the limits above exist although they could be extended real numbers (See *Note* (2.1.4)). The measure indicated above is referred to as the Lebesgue-Stieltjes measure on β.

To see how *Theorem* (4.2.19) can be used to determine probabilities associated with Borel sets B, note simply that the distribution function F of any random variable X satisfies the restrictions placed on the function g in the theorem and hence F induces a unique measure μ_F on the collection β of Borel sets of the real line as described in the theorem. We now consider the measure space $(\mathcal{R}, \beta, \mu_F)$.

Theorem 4.2.21

$$(\mathcal{R}, \beta, \mu_F) = (\mathcal{R}, \beta, P').$$

Proof: Let $a, b \in \mathcal{R}$.

$$\begin{aligned} P'[(a, b]] &= P[X^{-1}(a, b]] \\ &= P[X^{-1}(-\infty, b] - X^{-1}(-\infty, a]] \\ &= P[X^{-1}(-\infty, b]] - P[X^{-1}(-\infty, a]] \\ &= F(b) - F(a) \\ &= \mu_F((a, b]). \end{aligned}$$

$$\begin{aligned} P'[(-\infty, b]] &= F(b) = F(b) - 0 \\ &= F(b) - \lim_{x \to -\infty} F(x) = \mu_F((-\infty, b]). \end{aligned}$$

$$\begin{aligned} P'[(a, \infty)] &= 1 - P'[(-\infty, a]] \\ &= \lim_{x \to \infty} F(x) - F(a) \\ &= \mu_F((a, \infty)). \end{aligned}$$

$$\begin{aligned} P'[(-\infty, \infty)] &= 1 = 1 - 0 \\ &= \lim_{x \to \infty} F(x) - \lim_{x \to -\infty} F(x) = \mu_F((-\infty, \infty)). \end{aligned}$$

By the uniqueness of μ_F guaranteed in *Theorem* (4.2.19) we have that $P' = \mu_F$ as desired implying that

$(\mathcal{R}, \beta, \mu_F) = (\mathcal{R}, \beta, P')$.

The above theorem will now allow us to frame our question in terms of the more familiar Lebesgue-Stieltjes measure μ_F on the real line. That is, to determine $P[X \in B]$ we need simply to determine $\mu_F(B)$. Note however that this transition does not fully resolve the problem of actually obtaining a numerical value for

$$P[X \varepsilon B] = P'[B] = \mu_F(B).$$

The problem can be handled by considering the Lebesgue-Stieltjes integral.

Definition 4.2.22

Let $g : R \to R$ be a monotonic increasing function which is continuous from the right at each point and let Q be a non-negative Borel measurable function. We define the *Lebesgue-Stieltjes* integral of Q with respect to g denoted $\int Q\, d\, g$ by

$$\int Q\, d\, g = \int Q\, d\, \mu_g$$

where μ_g is the Lebesgue-Stieltjes measure on the Borel sets and the integral on the right is the abstract Lebesgue integral of Q provided the latter exists.

Note 4.2.23

If Q is not necessarily non-negative we can nevertheless define its Lebesgue-Stieltjes integral in a manner analogous to that used in defining the Lebesgue integral of such a function. Namely we define

$$\int Q\, d\, g = \int Q_+\, d\, g - \int Q_-\, d\, g$$

where Q_+ and Q_- are the positive and negative parts of Q, respectively, provided $\int Q_+\, d\, g < \infty$ and $\int Q_{-1}\, d\, g < \infty$

(See *Definition* (2.2.30) and *Definition* (2.2.32)).
We shall retain the convention of *Definition* (2.2.18) and denote the Lebesgue-Stieltjes integral of Q over B with respect to g by

$$\int_B Q \, d\, g = \int Q\, \chi_B \, d\, g$$

where χ_B is the indicator function defined in *Definition* (2.2.2).

Note 4.2.24

In order to apply the above definition to the situation at hand, namely that of calculating $P[X \in B]$ for some Borel set B, note first that the distribution function F of any random variable X satisfies the restrictions placed on the function g in *Definition* (4.2.22) and that the indicator function χ_B satisfies those placed on Q. Hence we can apply *Note* (4.2.23) and *Definition* (4.2.22) to obtain that

$$\int_B d\, F = \int \chi_B \, d\, F = \int \chi_B \, d\, \mu_F = \int_B d\, \mu_F = \mu_F(B) = P'[B] = P[X \in B].$$

Thus to evaluate $P[X \in B]$ we need only calculate the value of the Lebesgue-Stieltjes integral

$$\int_B d\, F$$

where F is the distribution function of the random variable X under consideration.

Although this problem is still rather formidable in general it is nevertheless quite simple in an extremely important case namely when the set B is of the form [a, b] for a and b real. This is a situation which constantly arises in dealing with random variables of the continuous type. In this instance we can draw on the following three theorems, presented without proof, which will relate the Lebesgue-Stieltjes integral to the more familiar Riemann-Stieltjes integral and finally to the Riemann integral of elementary calculus [5], [9].

*Theorem** 4.2.25

If g is a monotone function continuous on the right and h is Borel measurable, then

$$\int_{[a,b]} h \, d\, g = \int_a^b h \, d\, g$$

agrees with the Riemann-Stieltjes integral whenever the latter exists.

Note 4.2.26

Once again the distribution function F of any random variable X satisfies the conditions of the function g of *Theorem* (4.2.25) and $h(x) \equiv 1$ is Borel measurable, hence we may conclude that

$$P[X \varepsilon [a, b]] = \int_a^b d\, F_X = \int_a^b d\, F$$

where we now use the notation $d\, F_X$ to denote the Lebesgue-Stieltjes integral and $d\, F$ the Riemann-Stieltjes integral.

*Theorem** 4.2.27

If h is continuous on $[a, b]$ and g is increasing there, then

$$\int_a^b h \, d \, g$$

exists.

Note 4.2.28

Once again due to the fact that the distribution function F of any random variable is increasing and $h(x) \equiv 1$ is continuous, we have the condition of *Theorem* (4.2.27) satisfied and hence can conclude that the integral

$$\int_a^b d \, F$$

always exists. Thus in the case of a continuous random variable we can calculate $P[X \in [a, b]]$ by use of either the Lebesgue, Lebesgue-Stieltjes, or Riemann-Stieltjes integrals.

*Theorem** 4.2.29

If h and g' are continuous on $[a, b]$ then

$$\int_a^b h \, d \, g = \int_a^b h(x) \, d \, g(x) = \int_a^b h(x) \, g'(x) \, d \, x$$

where the integrals on the left are Riemann-Stieltjes integrals and the integral on the right is a Riemann integral.

Note 4.2.30

If we let $h(x) \equiv 1$, and $g(x) = F(x)$ and if X is a continuous random variable such that F' is continuous on $[a, b]$ then we may apply *Theorem* (4.2.29) and *Note* (4.2.12) to obtain that

$$P[X \in [a, b]] = \int_a^b dF = \int_a^b dF(x) = \int_a^b F'(x)\, dx = \int_a^b f(x)\, dx$$

where f is the density function of X. The condition that F' be continuous on $[a, b]$ is not restrictive as it is usually satisfied in the continuous case. Hence, in the case of a continuous random variable whose density f is known the problem of calculating $P[X \in [a, b]]$ is a simple problem in Riemann integration.

In the case of a discrete random variable X the density function f has a probabilistic interpretation when evaluated at a point x, namely,

$$f(x) = P[X = x].$$

This is not the situation in the continuous case as can be seen from the following theorem:

Theorem 4.2.31

Let X be a continuous random variable with density f. Then

$$P[X = x] = 0$$

for all $x \in \mathcal{R}$.

Proof: $P[X = x] = P[X \in [x, x]]$

$$= \int_x^x f(t)\, dt$$

$$= 0.$$

Examples 4.2.32

i) Let X be a random variable of the continuous type with uniform distribution over [0, 5]. Then

$$P[X \in [1, 3]] = \frac{2}{5} ;$$

ii) Let X be a random variable of the continuous type with exponential distribution with parameter $\alpha = 2$.

$$P[X < 1 \quad \text{or} \quad X > 3] = .617 ;$$

iii) For any continuous variable X with density f,

$$\int_{-\infty}^{\infty} f(t)\, d\, t = 1.$$

Proof:

i) $P[X \in [1, 3]] = \int_1^3 1/5 \; x\, dx = 2/5;$

ii)
$$\begin{aligned} P[X < 1 \quad \text{or} \quad X > 3] &= P[X \in ([0, 1) \cup (3, \infty))] \\ &= 1 - P[X \in [1, 3]] \\ &= 1 - \int_1^3 \tfrac{1}{2}\, e^{-x/2}\, dx \\ &= 1 + \int_1^3 - \tfrac{1}{2}\, e^{-x/2}\, dx \\ &= 1 + e^{-3/2} - e^{-\frac{1}{2}} \\ &= 1 - .383 \\ &= .617; \end{aligned}$$

iii) $\int_{-\infty}^{\infty} f(t)\, dt = \lim_{x\to\infty} \int_{-\infty}^{x} f(t)\, dt = \lim_{x\to\infty} F(x) = 1.$

4.3 *EXPECTATION AND MOMENT GENERATING FUNCTIONS*

In this section we shall use the theory of the Lebesgue integral introduced in Chapter Two to develop the concept and investigate some of the properties of *expectation*. Many of these properties hold in general for "*Lebesgue integrable functions*" but we prove them here in a probabilistic setting since this is the area of main concern in this text.

Definition 4.3.1

Let X be a random variable relative to $(\Omega, \mathcal{S}, P)$ integrable over Ω in the sense of Lebesgue. The *expectation (expected value)* of X, denoted $E[X]$, is defined by

$$E[X] = \int_{\Omega} X dP.$$

The following theorem provides a necessary and sufficient condition for the integrability of the random variable X and hence for the existence of an "*expectation*" for X. The property exemplified here is known in analysis as the property of "absolute integrability" of the Lebesgue integral.

Theorem 4.3.2

$E[X]$ exists if and only if $E[|X|]$ exists and furthermore

$$|E[X]| \leq E[|X|].$$

Proof: Note first that since X is a random variable and since the absolute value function is continuous by *Corollary* (4.1.17) $|X|$ is a random variable. Assume that $E[X]$ exists. By *Definition* (4.3.1),

$$-\infty < \int XdP < \infty.$$

By *Definition* (2.2.32), $\int XdP = \int X_+dP - \int X_-dP$. Thus X_+ and X_- are integrable. By *Theorem* (2.2.31) and *Theorem* (2.2.29),

$0 \le \int |X|dP = \int (X_+ + X_-)dP = \int X_+dP + \int X_-dP < \infty$. Thus $E[|X|]$ exists. Conversely, assume that $E[|X|]$ exists. Note that

$$0 \le X_+ \le |X| \quad \text{and} \quad 0 \le X_- \le |X|.$$

Thus by *Theorem* (2.2.20)

$$0 \le \int X_+dP \le \int |X|dP < \infty$$

and

$$0 \le \int X_-dP \le \int |X|dP < \infty.$$

By multiplying the second inequality by -1 and adding the two we obtain

$$-\infty < \int |X|dP \le \int X_+dP - \int X_-dP \le \int |X|dP < \infty.$$

Hence $E[X] = \int X_+dP - \int X_-dP$ exists

and $|E[X]| \le \int |X|dP = E[|X|]$.

Theorem 4.3.3

If X is a random variable, Y is a random variable such that $E[Y]$ exists and $|X| \leq |Y|$, then $E[X]$ exists and

$$E[|X|] \leq E[|Y|].$$

Proof: By *Theorem* (2.2.20), $0 \leq \int |X|dP \leq \int |Y|dP$. By *Theorem* (4.3.2), $\int |Y|dP$ exists. Hence $E[|X|] = \int |X|dP$ exists. By *Theorem* (4.3.2) $E[X]$ exists. Furthermore by the above inequality, $E[|X|] \leq E[|Y|]$.

Theorem 4.3.4

Let X be a random variable such that $E[X]$ exists and let c be a constant. Then

$$E[cX] = cE[X].$$

Proof: Assume that $c \geq 0$.
$E[cX] = \int cXdP = \int (cX)_+dP - \int (cX)_-dP.$

$$\text{Now } (cX)_+ = \sup\{0, cX\} = \begin{cases} 0 & \text{when } X < 0 \\ cX & \text{when } X \geq 0 \end{cases} = cX_+$$

and

$$(cX)_- = \sup\{0, -cX\} = \begin{cases} 0 & \text{when } X \geq 0 \\ -cX & \text{when } X < 0 \end{cases} = cX_-.$$

$$\begin{aligned} \text{Thus } E[cX] &= \int cX_+dP - \int cX_-dP \\ &= c\int X_+dP - c\int X_-dP \quad \text{by } \textit{Theorem } (2.2.28). \end{aligned}$$

$$\begin{aligned} E[cX] &= c[\int X_+dP - \int X_-dP] \\ &= cE[X]. \end{aligned}$$

Assume $c < 0$. Then

$$(cX)_+ = \sup\{0, cX\} = \begin{cases} 0 & \text{when } X \geq 0 \\ cX & \text{when } X < 0 \end{cases} = -cX_-$$

and

$$(cX)_- = \sup\{0, -cX\} = \begin{cases} 0 & \text{when } X < 0 \\ -cX & \text{when } X \geq 0 \end{cases} = -cX_+.$$

Thus $E[cX] = \int cXdP = \int (cX)_+dP - \int (cX)_-dP$

$$= \int -cX_-dP - \int -cX_+dP.$$

Since $-c > 0$, by *Theorem* (2.2.28),

$$\begin{aligned} E[cX] &= -c \int X_-dP + c \int X_+dP \\ &= c \left[\int X_+dP - \int X_-dP\right] \\ &= cE[X]. \end{aligned}$$

Corollary 4.3.5

Let c be a constant. Then

$$E[c] = c.$$

Proof: Let $X(\omega) \equiv 1$. Consider $E[X] = \int 1dP = P[\Omega] = 1$ by *Example* (2.2.8) and the second axiom of probability. Thus $E[1]$ exists and has value 1. $E[c] = E[c \cdot 1] = cE[1] = c$ by *Theorem* (4.3.4).

Theorem 4.3.6

Let X and Y be random variables such that $E[X]$ and $E[Y]$ exist. Then $E[X + Y]$ exists and

$$E[X + Y] = E[X] + E[Y].$$

Proof: By *Theorem* (4.3.2), $E[|X|]$ and $E[|Y|]$ exist. Note that by *Theorem* (2.2.29)
$\int(|X| + |Y|)dP = \int |X|dP + \int|Y|dP = E[|X|] + E[|Y|]$ exists. Thus since $|X + Y| \leq |X| + |Y|$ we may apply *Theorem* (4.3.3) to conclude that $E[X + Y]$ exists. Since X_+, Y_+, X_-, Y_- are all non-negative we may apply *Theorem* (2.2.34) to obtain that

$$\int(X + Y)dP = \int (X_+ + Y_+)dP - \int (X_- + Y_-)dP.$$

By *Theorem* (2.2.29) we obtain

$$\int (X + Y)dP = \int X_+dP + \int Y_+dP - \int X_-dP - \int Y_-dP.$$

Rearranging terms and applying *Definition* (2.2.32) we obtain

$$\begin{aligned}\int (X + Y)dP &= [\int X_+dP - \int X_-dP] + [\int Y_+dP - \int Y_-dP] \\ &= E[X] + E[Y].\end{aligned}$$

In order to obtain several other interesting results pertaining to expectation we shall need the following definition which is the probabilistic analog of the measure theoretic concept of a proposition or statement S holding "*almost everywhere.*" This concept will also be of use in our later discussion of modes of convergence.

Definition 4.3.7

Let $(\Omega, \mathcal{F}, P)$ be a probability space and let S be a proposition relative to this space. S is said to hold *with probability one* (P-almost everywhere-P.a.e.) if there exists a subset $A \in \mathcal{F}$ such that $P[A] = 0$ and S holds on the complement of A.

Note 4.3.8

The above definition is often stated in the following equivalent form: "Let $(\Omega, \mathcal{F}, P)$ be a probability space and let S be a proposition relative to this space. Let $\{\omega : S \text{ fails}\} \in \mathcal{F}$. S is said to hold with probability one if and only if

$$P\{\omega : S \text{ fails}\} = 0".$$

Note also that if $(\Omega, \mathcal{F}, P)$ is complete (See *Definition* 3.3.1)) then the condition that $\{\omega : S \text{ fails}\} \in \mathcal{F}$ is trivially satisfied and need not be stated explicitly. However, if the completeness assumption is not made then this requirement is essential to the equivalency of the two definitions.

Verification: Let $(\Omega, \mathcal{F}, P)$ be a probability space and let S be a proposition relative to this space. Let $\{\omega : S \text{ fails}\} \in \mathcal{F}$. Assume that $P\{\omega : S \text{ fails}\} = 0$. Then simply take A of *Definition* (4.3.7) to be $\{\omega : S \text{ fails}\}$. Now assume that S holds with probability one. Then there exists a set $A \in \mathcal{F}$ such that $P[A] = 0$ and S holds on the complement of A. Thus $\{\omega : S \text{ fails}\} \subseteq A$. Since $\{\omega : S \text{ fails}\} \in \mathcal{F}$, by *Theorem* (3.1.5) $0 \leq P\{\omega : S \text{ fails}\} \leq P[A] = 0$. That is, $P\{\omega : S \text{ fails}\} = 0$. If $(\Omega, \mathcal{F}, P)$ is complete, then the fact that $\{\omega : S \text{ fails}\} \subseteq A$ implies by *Definition* (3.3.1) that $\{\omega : S \text{ fails}\} \in \mathcal{F}$ and it is therefore not necessary to state this fact explicitly.

Theorem 4.3.9

Let X be a random variable such that $X = 0$ with

probability one. Then $E[X]$ exists and has value 0.

Proof: By *Theorem* (4.3.2), $E[X]$ exists if and only if $E[|X|]$ exists. By *Note* (4.3.8), $\{\omega : X(\omega) \neq 0\} \in \mathcal{F}$ and $P\{\omega : X(\omega) \neq 0\} = 0$. Note that
$E = \{\omega : |X(\omega)| > 0\} = \{\omega : X(\omega) \neq 0\}$ and hence $P[E] = 0$.
By *Exercise* (2.2.1)

$$\int |X| dP = 0$$

implying that $E[|X|]$ exists and has value 0.
By *Theorem* (4.3.2), $0 \leq |E[X]| \leq E[|X|] = 0$ implying that $E[X] = 0$ as desired.

Theorem 4.3.10

Let X be a random variable and let $E_1, E_2, E_3 \ldots$ be a sequence of events which partitions Ω.
$\int XdP$ exists if and only if for each i

$$\int_{E_i} XdP \text{ exists and } \sum_{i=1}^{\infty} \int_{E_i} |X| dP < \infty.$$

In this case $\displaystyle\sum_{i=1}^{\infty} \int_{E_i} X\, dP = \int XdP.$

Proof: Assume that $E_1, E_2, E_3, \ldots$ is a sequence of events which partitions Ω and that $\int_{E_i} X\, dP$ exists for each i and

$$\sum_{i=1}^{\infty} \int_{E_i} |X| dP < \infty.$$ Then we have

$$\sum_{i=1}^{\infty} \int_{E_i} |X| dP = \sum_{i=1}^{\infty} [\int_{E_i} X_+ dP + \int_{E_i} X_- dP]$$

$$= \sum_{i=1}^{\infty} \int_{E_i} X_+ dP + \sum_{i=1}^{\infty} \int X_- dP$$

since both series are positive termed.

Thus

$$\sum_{i=1}^{\infty} \int_{E_i} X_+ dP < \infty \quad \text{and} \quad \sum_{i=1}^{\infty} \int_{E_i} X_- dP < \infty.$$

Let $X_n^+ = \sum_{i=1}^{n} X_+ \chi_{E_i}$. Then

$$\lim_{n\to\infty} X_n^+ = X_+ \quad \text{and} \quad 0 \le X_n^+ \le X_{n+1}^+.$$

By the Monotone Convergence Theorem, *Theorem* (2.2.26)

$\lim_{n\to\infty} \int X_n^+ \, dP = \int X_+ dP$. But

$$\lim_{n\to\infty} \int X_n^+ \, dP = \lim_{n\to\infty} \int \sum_{i=1}^{n} X_+ \chi_{E_i} dP$$

$$= \lim_{n\to\infty} \sum_{i=1}^{n} \int X_+ \chi_{E_i} dP$$

$$= \lim_{n\to\infty} \sum_{i=1}^{n} \int_{E_i} X_+ dP = \sum_{i=1}^{\infty} \int_{E_i} X_+ \, dP < \infty.$$

Similarly $\int X_- \, dP = \sum_{i=1}^{\infty} \int_{E_i} X_- dP < \infty.$

Thus,

$$\int X dP = \int X_+ \, dP - \int X_- \, dP \quad \text{exists.}$$

Furthermore

$$\int XdP = \sum_{i=1}^{\infty} \int_{E_i} X_+dP - \sum_{i=1}^{\infty} \int_{E_i} X_-dP$$

$$= \sum_{i=1}^{\infty} [\int_{E_i} X_+dP - \int_{E_i} X_-dP]$$

$$= \sum_{i=1}^{\infty} \int_{E_i} XdP.$$

To prove the converse, assume that $\int XdP$ exists. Note that $\int XdP = E[X]$ and hence by *Theorem* (4.3.2), $E[|X|] = \int |X|\, dP$ exists. That is,

$$\int_{\bigcup_{i=1}^{\infty} E_i} |X|\, dP < \infty. \text{ By}$$

Exercise (2.2.3),

$$\int_{\bigcup_{i=1}^{\infty} E_i} |X|\, dP = \sum_{i=1}^{\infty} \int_{E_i} |X| dP < \infty.$$

Hence for each i, $\int_{E_i} |X|\, dP = E[|X|\, \chi_{E_i}]$ exists.

By *Theorem* (4.3.2)

$$E[X\, \chi_{E_i}] = \int X\, \chi_{E_i}\, dP = \int_{E_i} XdP$$

exists for each i and the proof is complete.

Theorem 4.3.11

Let X be a random variable such that $E[X]$ exists and let Y be a random variable such that $X = Y$ with probability one. Then $E[Y]$ exists and $E[X] = E[Y]$.

Proof: We shall first show that $E[Y]$ exists. By *Definition* (4.3.7) there exists an element $A \in \mathcal{A}$ such that $P[A] = 0$ and $X(\omega) = Y(\omega)$ for $\omega \in A'$. Let $E_1 = A$, $E_2 = A'$, $E_i = \phi$ for $i \geq 3$. By *Definition* (4.3.1), *Theorem* (4.3.2), and *Theorem* (4.3.10),

$$\int |X|dP = \sum_{i=1}^{\infty} \int_{E_i} |X|dP \quad \text{exists.}$$

However $\sum_{i=1}^{\infty} \int_{E_i} |X|dP = \int_{A'} |X|dP$ by *Exercise* (2.2.2).

Similarly $E[|Y|]$ if it exists is given by $\int_{A'} |Y|dP$.

However since

$X(\omega) = Y(\omega)$ for $\omega \in A'$, we have

$$\int_{A'} |Y|dP = \int_{A'} |X|dP.$$

Hence $E[|Y|]$ exists and by *Theorem* (4.3.2), $E[Y]$ exists. Note that $X - Y = 0$ with probability one. By *Theorem* (4.3.9) $E[X - Y]$ exists and $E[X - Y] = 0$. By *Theorem* (4.3.4) and *Theorem* (4.3.6) $E[X - Y] = E[X] - E[Y] = 0$. Hence $E[X] = E[Y]$ as desired.

Thus far our emphasis has been solely upon the theoretical properties of the concept of expectation. As was the case when discussing abstract probabilities it will be of interest to see how the expectation of a random variable can be calculated in the discrete and continuous cases. It will be seen once again that the problem can be reduced to one of summation in the discrete case and of Riemann integration in the continuous case.

Theorem 4.3.12

Let $(\Omega, \mathcal{F}, P)$ be a probability space and X a random variable of the discrete type with range $\{x_\gamma : \gamma \in \Gamma\}$ and density f. Let $g : R \to R$ be Borel measurable. $E[|g(X)|]$ exists if and only if

$$\sum_{\gamma \in \Gamma} |g(x_\gamma)| f(x_\gamma) < \infty.$$

Proof: Assume that $E[|g(X)|]$ exists. Then by *Definition* (4.3.1) we have that

$$\int |g(X)| \, dP \quad \text{exists.}$$

Let $E_\gamma = \{\omega : X(\omega) = x_\gamma\}$ and note that $\{E_\gamma : \gamma \in \Gamma\}$ is a partition of Ω. Hence

$$\begin{aligned} \int |g(X)| dP &= \bigcup_{\gamma \in \Gamma} \int_{E_\gamma} |g(X)| dP \\ &= \sum_{\gamma \in \Gamma} \int_{E_\gamma} |g(X)| dP \\ &= \sum_{\gamma \in \Gamma} \int |g(X)| \chi_{E_\gamma} dP. \end{aligned}$$

Note that $|g(X)|\chi_{E_\gamma}(\omega) = \begin{cases} |g(x_\gamma)| & \text{if } \omega \in E_\gamma \\ 0 & \text{otherwise.} \end{cases}$

Hence by *Definition* (2.2.7) we have that

$$\begin{aligned} \int |g(X)| dP &= \sum_{\gamma \in \Gamma} |g(x_\gamma)| \; P[E_\gamma] \\ &= \sum_{\gamma \in \Gamma} |g(x_\gamma)| \; P[X = x_\gamma] \end{aligned}$$

$$= \sum_{\gamma \varepsilon \Gamma} |g(x_\gamma)| f(x_\gamma).$$

Thus if $E[|g(X)|]$ exists then $\sum_{\gamma \varepsilon \Gamma} |g(x_\gamma)| f(x_\gamma)$ exists as was desired.

To prove the converse, assume that $\sum_{\gamma \varepsilon \Gamma} |g(x_\gamma)| f(x_\gamma) < \infty$. Note that for each $\gamma \varepsilon \Gamma$,

$$\int_{E_\gamma} |g(X)| dP = \int_{E_\gamma} |g(x_\gamma)| dP$$
$$= |g(x_\gamma)| P[E_\gamma]$$
$$= |g(x_\gamma)| f(x_\gamma) < \infty.$$

Thus we may apply *Theorem* (4.3.10) to obtain that $\int |g(X)| dP = E[|g(X)|]$ exists.

Theorem 4.3.13

Let $(\Omega, \mathcal{F}, P)$ be a probability space and X a random variable of the discrete type with range $\{\mathbf{x}_\gamma : \gamma \varepsilon \Gamma\}$ and density f. Let $g : R \to R$ be Borel measurable. $E[g(X)]$ exists if and only if

$$\sum_{\gamma \varepsilon \Gamma} |g(x_\gamma)| f(x_\gamma) < \infty$$

and in this case

$$E[g(X)] = \sum_{\gamma \varepsilon \Gamma} g(x_\gamma) f(x_\gamma).$$

Proof: *Theorem* (4.3.12) implies that $E[|g(X)|]$ exists if and only if $\sum_{\gamma \varepsilon \Gamma} |g(x_\gamma)| f(x_\gamma) < \infty$. *Theorem* (4.3.2) implies that $E[g(X)]$ exists if and only if $E[|g(X)|]$

exists. Combining these two results we obtain that $E[g(X)]$ exists if and only if $\sum_{\gamma \in \Gamma} |g(x_\gamma)| f(x_\gamma) < \infty$. The fact that $E[g(X)] = \sum_{\gamma \in \Gamma} g(x_\gamma) f(x_\gamma)$ follows from *Theorem* (4.3.10) and the fact that for each $\gamma \in \Gamma$

$$\begin{aligned}\int_{E_\gamma} g(X)\,dP &= \int_{E_\gamma} g(x_\gamma)\,dP \\ &= g(x_\gamma) \int_{E_\gamma} dP \\ &= g(x_\gamma) P[E_\gamma] \\ &= g(x_\gamma) f(x_\gamma).\end{aligned}$$

Note 4.3.14

The theorem stated above is fairly general in that the function g is not completely specified. Thus this theorem in essence allows us to find the expectation of some functional form of X by using the knowledge of the density function of the base variable X. It does not require us to first find the density for the random variable $g(X)$. The result obtained is of considerable interest in applied situations. It is often used as the definition of expectation in the discrete case in more elementary courses in probability. Note also that if g is the identity map then we obtain a computational formula for $E[X]$.

Examples 4.3.15

i) Let X be a discrete random variable with a point binomial distribution with parameter p. $E[X] = p$;

ii) Let X be a discrete random variable with a binomial distribution with parameters n and p.

$E[X] = np$ and

$E[X^2] = n^2p^2 - np^2 + np$;

Hint: Use the *Binomial Theorem, Theorem* (2.4.1).

Note also that

$$x^2 = x(x - 1) + x.$$

iii) Let X be a discrete random variable with a Poisson distribution with parameter s.

$E[X] = s$ and

$E[X^2] = s^2 + s$;

Verification:

i) $E[X] = 1 \cdot p + 0(1 - p) = p$

ii) $$E[X] = \sum_{x=0}^{n} x\binom{n}{x}p^x(1 - p)^{n-x}$$

$$= \sum_{x=1}^{n} x\binom{n}{x}p^x(1 - p)^{n-x}$$

$$= \sum_{x=1}^{n} \frac{xn(n - 1)!}{x(x - 1)!(n - x)!} p^x(1-p)^{n-x} .$$

Let $y = x - 1$. Substituting we obtain

$$E[X] = \sum_{y=0}^{n-1} \frac{n(n - 1)!}{y!(n-1-y)!} p^{y+1}(1-p)^{n-y-1}$$

$$= np \sum_{y=0}^{n-1} \binom{n-1}{y} p^y(1-p)^{n-y-1} .$$

By the *Binomial Theorem, Theorem* (2.4.1),

$$\sum_{y=0}^{n-1} \binom{n-1}{y}p^y(1-p)^{n-y-1} = (p + (1-p))^{n-1} = 1.$$

Hence $E[X] = n\,p$.

$$E[X^2] = \sum_{x=0}^{n} x^2\binom{n}{x}p^x(1-p)^{n-x}$$

$$= \sum_{x=0}^{n} [x(x-1) + x] \binom{n}{x} p^x (1-p)^{n-x}$$

$$= \sum_{x=0}^{n} x(x-1) \binom{n}{x} p^x (1-p)^{n-x} + \sum_{x=0}^{n} x \binom{n}{x} p^x (1-p)^{n-x}$$

$$= \sum_{x=0}^{n} \frac{x(x-1)n(n-1)(n-2)!}{x(x-1)(x-2)!(n-x)!} p^x (1-p)^{n-x} + E[X]$$

$$= \sum_{x=2}^{n} \frac{n(n-1)(n-2)!}{(x-2)!(n-x)!} p^x (1-p)^{n-x} + E[X] .$$

Let $y = x-2$.

$$E[X^2] = \sum_{y=0}^{n-2} \frac{n(n-1)(n-2)!}{y!(n-y-2)!} p^{y+2} (1-p)^{n-y-2} + E[X]$$

$$= n(n-1)p^2 \sum_{y=0}^{n-2} \binom{n-2}{y} p^y (1-p)^{n-2-y} + E[X] .$$

Again applying the *Binomial Theorem*, we obtain

$$E[X^2] = n(n-1)p^2 + np = n^2p^2 - np^2 + np.$$

iii) $$E[X] = \sum_{x=0}^{\infty} \frac{xe^{-s}s^x}{x!}$$

$$= \sum_{x=1}^{\infty} \frac{xe^{-s}s^x}{x(x-1)!}$$

$$= \sum_{x=1}^{\infty} \frac{e^{-s}s^x}{(x-1)!} .$$

Let $y=x-1$. Then

$$E[X] = \sum_{y=0}^{\infty} \frac{e^{-s}s^{y+1}}{y!}$$

$$= e^{-s}s \sum_{y=0}^{\infty} \frac{s^y}{y!} .$$

Note that by *Theorem* (2.4.4), $\sum_{y=0}^{\infty} \frac{s^y}{y!}$ is the Maclaurin Series expansion for e^s and hence we have that

$$E[X] = e^{-s}se^{s} = s \quad \text{as desired.}$$

$$E[X^2] = \sum_{x=0}^{\infty} \frac{x^2 e^{-s} s^x}{x!}$$

$$= \sum_{x=0}^{\infty} [x(x-1) + x] \frac{e^{-s} s^x}{x!}$$

$$= \sum_{x=0}^{\infty} \frac{x(x-1)e^{-s} s^x}{x!} + E[X]$$

$$= \sum_{x=2}^{\infty} \frac{x(x-1)e^{-s} s^x}{x(x-1)(x-2)!} + E[X].$$

Let $y = x - 2$. Then

$$E[X^2] = \sum_{y=0}^{\infty} \frac{e^{-s} s^{y+2}}{y!} + E[X]$$

$$= s^2 e^{-s} \sum_{y=0}^{\infty} \frac{s^y}{y!} + E[X]$$

$$= s^2 + s.$$

In the case of a continuous random variable we shall be forced once again to consider the Lebesgue-Stieltjes, Riemann-Stieltjes, and Riemann integrals in order to compute expectations. The key to the problem is found in the following theorem from Kingman and Taylor, [4].

Theorem *4.3.16

Let $(\Omega, \mathcal{A}, P)$ be a probability space and let X be a random variable defined on Ω with distribution function F_X. Let $g : R \to R$ be Borel measurable. Then

$$\int g(X)\,dP = \int g\,dF_X = \int g(x)\,dF_X(x)$$

where the integral on the left is a Lebesgue integral with respect to $(\Omega, \mathcal{A}, P)$ and the two on the right are the Lebesgue-Stieltjes integrals of g with respect to F_X. These are considered equal in the sense that if either exists then so does the other and they are equal.

The following theorem gives necessary and sufficient conditions for the existence of $E[|g(X)|]$ in the case where $g : R \to R$ is continuous and X is a continuous random variable such that its density f has at most a finite number of discontinuities. These restrictions will allow us to make the transition between the Lebesgue and Riemann integrals by use of theorems previously considered. In actuality these restrictions are not at all stringent as most random variables of the continuous type of interest to practitioners satisfy these requirements. This theorem will eventually allow us to consider random variables of the form X^n, $(X - c)^n$, and e^{cX} where $n \in N$ and c is a constant. These functional forms are of particular interest to statisticians.

Theorem 4.3.17

Let $(\Omega, \mathcal{A}, P)$ be a probability space and let X be a random variable of the continuous type with density f and distribution function F_X. Assume that f has at most a finite number of discontinuities. Let $g : R \to R$ be continuous.

$E[|g(X)|]$ exists if and only if

$$\int_{-\infty}^{\infty} |g(x)| f(x)\,dx < \infty$$

where the integral involved is a Riemann integral.

Proof: Let X, f, F_X and g be as described in the theorem and let $\{x_1, x_2, \ldots, x_n\}$ be the points of discontinuity of f where $x_1 < x_2 < x_3 < \cdots < x_n$. Assume that $E[|g(X)|]$ exists. Then by *Definition* (4.3.1),

$$E[|g(X)|] = \int |g(X)|\,dP.$$

By *Theorem* (2.1.42), *Theorem* (4.3.16), *Theorem* (4.2.25), *Note* (4.2.26) and *Theorem* (4.2.27) we have

$$\begin{aligned}\int |g(X)|\,dP &= \int_{-\infty}^{\infty} |g(x)|\,dF_X(x)\\ &= \lim_{a\to-\infty} \int_a^{x_1} |g(x)|\,dF_X(x) + \int_{x_1}^{x_2} |g(x)|\,dF_X(x) + \cdots\\ &\quad \cdots \lim_{b\to\infty} \int_{x_n}^{b} |g(x)|\,dF_X(x)\end{aligned}$$

$$= \lim_{a\to-\infty} \int_a^{x_1} |g(x)|\,dF(x) + \int_{x_1}^{x_2} |g(x)|\,dF(x) + \cdots \lim_{b\to\infty} \int_{x_n}^{b} |g(x)|\,dF(x).$$

However by *Theorem* (4.2.29) we have that

$$\int |g(x)|dP = \lim_{a\to-\infty} \int_a^{x_1} |g(x)|f(x)dx + \int_{x_1}^{x_2} |g(x)|f(x)dx + \cdots$$

$$\lim_{b\to\infty} \int_{x_n}^{b} |g(x)|f(x)dx$$

$$= \int_{-\infty}^{\infty} |g(x)|f(x)dx < \infty.$$

The converse may be proved by reversing the argument.

Theorem 4.3.18

Let $(\Omega, \mathfrak{f}, P)$ be a probability space and let X be a random variable of the continuous type with density f and distribution function F_X. Assume that f has at most a finite number of discontinuities. Let $g : R \to R$ be continuous. Then $E[g(X)]$ exists if and only if

$$\int_{-\infty}^{\infty} |g(x)|f(x)dx < \infty$$

where the integral involved is a Riemann integral and in this case

$$E[g(X)] = \int_{-\infty}^{\infty} g(x)f(x)dx.$$

Proof: By *Theorem* (4.3.17) $E[|g(X)|]$ exists if and only if $\int_{-\infty}^{\infty} |g(x)|f(x)dx < \infty$. By *Theorem* (4.3.2), $E[g(X)]$ exists if and only if $E[|g(X)|]$ exists. Hence $E[g(X)]$ exists if and only if the given integral is finite. We may use *Theorem* (4.3.16) and an argument analogous to that given in the proof of *Theorem* (4.3.17) to verify that $E[g(X)] = \int_{-\infty}^{\infty} g(x)f(x)dx.$

Note 4.3.19

Just as in the discrete case, the above theorem allows us to calculate the expectation of certain functional forms of X by using the density of the base variable X. It also is sometimes taken as the definition of expectation in the case of a continuous random variable.

Example 4.3.20

i) Let X be a random variable of the continuous type with uniform distribution over [0, 5].

$$E[X] = \frac{25}{10} \quad \text{and} \quad E[X^2] = \frac{25}{3};$$

ii) Let X be a random variable of the continuous type with exponential distribution with parameter $\alpha = 2$. $E[X] = 2$ and $E[X^2] = 8$;

iii) Let X be a random variable of the continuous type with Cauchy distribution. Show that the Cauchy principal value of

$$\int_{-\infty}^{\infty} x \frac{1}{\pi(1+x^2)} dx$$

exists but $E[X]$ does not exist (See *Definition* (2.4.2) and *Note* (2.4.3)).

Verification:

i) $$E[X] = \int_{-\infty}^{\infty} x\, 1/5\, dx = 1/5\, x^2/2 \Big|_0^5 = \frac{25}{10}$$

$$E[X^2] = 1/5 \int_0^5 x^2 dx = 1/5 \frac{x^3}{3} \Big|_0^5 = \frac{25}{3}$$

ii) $$E[X] = \int_0^{\infty} 1/2\, xe^{-x/2} dx.$$ Let $Z = \frac{x}{2}$. Then

$$E[X] = \int_0^{\infty} \tfrac{1}{2}(2Z)e^{-Z} 2dZ$$

$$= 2 \int_0^\infty Ze^{-Z} dZ$$

$$= 2\ \Gamma(2) = 2$$ by *Definition* (2.4.5) and *Theorem* (2.4.6).

$$E[X^2] = \int_0^\infty \tfrac{1}{2}\ x^2 e^{-x/2} dx.$$

Let $Z = x/2$. Then

$$E[X^2] = \int_0^\infty \tfrac{1}{2}(4Z^2) e^{-Z} 2\ dZ$$

$$= 4 \int_0^\infty Z^2 e^{-Z}\ dZ$$

$$= 4\ \Gamma(3) = 8.$$

iii) $$\int_{-\infty}^\infty x \frac{1}{\Pi(1+x^2)}\ dx = \frac{1}{\Pi} \int_{-\infty}^\infty \frac{x}{1+x^2}\ dx$$

$$= \lim_{h\to\infty} \left[\frac{1}{2\Pi} \int_{-h}^{h} \frac{2x}{1+x^2}\ dx\right]$$

$$= \lim_{h\to\infty} \left[\frac{1}{2\Pi}\ \ell n(1+x^2)\Big|_{-h}^{h}\ \right]$$

$$= \lim_{h\to\infty} \left[\frac{1}{2\Pi}\ (\ell n(1+h^2) - \ell n(1+h^2)\right]$$

$$= 0.$$

Hence the Cauchy principal value exists. However

$$E[|X|] = \int_{-\infty}^\infty |x| \frac{1}{\Pi(1+x^2)}\ dx$$

$$= \frac{1}{\Pi} \int_{-\infty}^0 |x|\ \frac{1}{1+x^2}\ dx + \frac{1}{\Pi} \int_0^\infty |x|\ \frac{1}{1+x^2}\ dx$$

$$= \frac{1}{\Pi} \int_{-\infty}^{0} -x \frac{1}{1+x^2} dx + \frac{1}{\Pi} \int_{0}^{\infty} \frac{x}{1+x^2} dx$$

$$= -\frac{1}{2\Pi} \lim_{h\to-\infty} \int_{h}^{0} \frac{2x}{1+x^2} dx + \frac{1}{2\Pi} \lim_{h\to\infty} \int_{0}^{h} \frac{2x}{1+x^2} dx$$

$$= -\frac{1}{\Pi} \lim_{h\to-\infty} \ell n(1+x^2)\Big|_{h}^{0} + \frac{1}{2\Pi} \lim_{h\to\infty} \ell n(1+x^2)\Big|_{0}^{h}$$

$$= -\frac{1}{2\Pi} \lim_{h\to-\infty} [\ell n 1 - \ell n(1+h^2)] +$$

$$\frac{1}{2\Pi} \lim_{h\to\infty} [\ell n(1+h^2) - \ell n 1]$$

$$= \frac{1}{2\Pi} \cdot \lim_{h\to-\infty} \ell n(1+h^2) + \frac{1}{2\Pi} \lim_{h\to\infty} \ell n(1+h^2)$$

$$= \infty + \infty = \infty.$$

Thus $E[|X|]$ does not exist and by *Theorem* (4.3.2) $E[g(X)]$ does not exist.

We have mentioned previously that there are several functional forms of random variables which are of particular interest to statisticians namely those of the form X^n, $(X-c)^n$ and e^{cX} where $n \in N$ and c is real. These random variables allow us to define the concepts of nth moments, n^{th} central moments, and moment generating function for the random variable X, respectively.

Definition 4.3.21

Let $(\Omega, \mathcal{F}, P)$ be a probability space and X a random variable defined on Ω. Let $n \in N$. The nth moment or n^{th} ordinary moment of X, denoted μ_n, is defined by

$$\mu_n = E[X^n]$$

provided this expectation exists.

Note 4.3.22

The first ordinary moment μ_1 is used extensively in practice and is generally denoted simply by μ and referred to as the mean of X.

Definition 4.3.23

Let $(\Omega, \mathcal{J}, P)$ be a probability space and X a random variable defined on Ω. Let $n \in N$. The nth central moment of X, denoted η_n, is defined by

$$\eta_n = E[(X-\mu)^n]$$

provided this expectation exists.

Note 4.3.24

The second central moment η_2 is also used extensively in practice and is generally denoted by σ^2 or Var (X). σ^2 is referred to as the *variance* of X and $\sigma = \sqrt{\sigma^2}$ is called the *standard deviation* of X. σ^2 is used as a measure of variability.

In the case of discrete or continuous random variables X, we can apply *Theorem* (4.3.13) or *Theorem* (4.3.18) to calculate the values of μ and σ^2. However, the following theorem provides a computational tool for calculating σ^2. This result is useful in that it allows one to work solely with ordinary moments which are computationally easier to handle than are central moments.

Theorem 4.3.25

$$\sigma^2 = \mu_2 - \mu^2$$

Proof: $\sigma^2 = E[(X-\mu)^2]$. Expanding the right hand side we obtain $\sigma^2 = E[X^2 - 2\mu X + \mu^2]$. Applying *Theorem* (4.3.6), *Corollary* (4.3.5) and *Theorem* (4.3.4) we have

$$\sigma^2 = E[X^2] - 2\mu E[X] + \mu^2.$$

By *Definition* (4.3.21), $\sigma^2 = \mu_2 - \mu^2$.

Examples 4.3.26

i) Let X be a random variable of the discrete type with a binomial distribution with parameters n and p.

$$\sigma^2 = n(p)(1-p);$$

ii) Let X be a random variable of the discrete type with a Poisson distribution with parameter s. $\sigma^2 = s$;

iii) Let X be a random variable of the continuous type with exponential distribution with parameter 2. $\sigma^2 = 4$.

Definition 4.3.27

Let $(\Omega, \mathcal{F}, P)$ be a probability space and let X be a random variable defined on Ω. The function $m_X(\theta)$ defined by

$$m_X(\theta) = E[e^{\theta X}]$$

is called the *moment generating function* of X provided there exists a positive number h such that $m_X(\theta)$ exists for $\theta \in (-h, h)$.

Examples 4.3.28

i) Let X be a random variable of the discrete type with binomial distribution with parameters n and p. $m_X(\theta) = [pe^{\theta} + (1-p)]^n$;

ii) Let X be a random variable of the discrete type with Poisson distribution with parameter s.

$m_X(\theta) = e^{s(e^{\theta}-1)}$;

iii) Let X be a random variable of the continuous type with exponential distribution with parameter α.

$m_X(\theta) = (1 - \alpha\theta)^{-1}$;

iv) Let X be a random variable of the continuous type with uniform distribution over [a, b].

$$m_X(\theta) = \begin{cases} \frac{1}{\theta(b-a)} [e^{\theta b} - e^{\theta a}], & \theta \neq 0 \\ 1, & \theta = 0. \end{cases}$$

Verification:

i) $m_X(\theta) = E[e^{\theta X}] = \sum_{x=0}^{n} e^{\theta x} \binom{n}{x} p^x (1-p)^{n-x}$

$$= \sum_{x=0}^{n} \binom{n}{x} (pe^{\theta})^x (1-p)^{n-x}$$

$= [pe^{\theta} + (1-p)]^n$ by *Theorem* (2.4.1).

ii) $$m_X(\theta) = \sum_{x=0}^{\infty} \frac{e^{\theta x} e^{-s} s^x}{x!}$$

$$= e^{-s} \sum_{x=0}^{\infty} \frac{(se^{\theta})^x}{x!}$$

$$= e^{se^{\theta}} e^{-s}$$

$$= e^{s(e^{\theta} - 1)} \quad \text{by } \textit{Theorem } (2.4.4).$$

iii) Without loss of generality we may assume that $|\theta| < 1/\alpha$ since we are only concerned with the existence of $E[e^{\theta X}]$ on some open interval about 0. Thus

$$m_X(\theta) = \int_0^{\infty} \frac{1}{\alpha} e^{\theta x} e^{-x/\alpha} dx$$

$$= \frac{1}{\alpha} \int_0^{\infty} e^{(\theta - \frac{1}{\alpha})x} dx$$

$$= \frac{1}{(\theta - \frac{1}{\alpha})\alpha} \int_0^{\infty} (\theta - \frac{1}{\alpha}) e^{(\theta - \frac{1}{\alpha})x} dx$$

$$= \frac{1}{\alpha\theta - 1} e^{(\theta - \frac{1}{\alpha})x} \Big|_0^{\infty}$$

$$= \frac{1}{(\alpha\theta - 1)} [0-1] \quad \text{since} \quad \theta - \frac{1}{\alpha} < 0.$$

Thus $m_X(\theta) = \frac{1}{(1-\alpha\theta)} = (1 - \alpha\theta)^{-1}$.

iv) $$m_X(\theta) = E[e^{\theta X}] = \int_a^b \frac{1}{b-a} e^{\theta x} dx$$

$$= \frac{1}{(b-a)\theta} e^{\theta x} \Big|_a^b$$

$$= \frac{1}{\theta(b-a)} [e^{\theta b} - e^{\theta a}] \quad \theta \neq 0.$$

$$m_X(0) = E[e^{0X}] = E[1] = 1.$$

The terminology used in *Definition* (4.3.27) is descriptive in that the function $m_X(\theta)$ is used to do just what the wording implies, namely, generate ordinary moments of X when they exist.

Theorem 4.3.29

If the moment generating function of X, $m_X(\theta)$, exists for $-h<\theta<h$, then

$$\mu_n = E[X^n]$$

exists for all natural numbers.

Proof: Let $a = \frac{h}{2} > 0$. Then for $x \in (-\infty, \infty)$ we have

$$\frac{|ax|^n}{n!} \leq 1 + |ax| + \frac{|ax|^2}{2!} + \cdots = e^{|ax|}.$$

Let $F(x)$ be the distribution function of the random variable X. Then

$$\int_{-\infty}^{\infty} \frac{|ax|^n}{n!} dF(x) \leq \int_{-\infty}^{\infty} e^{|ax|} dF(x)$$

$$= \int_{-\infty}^{0} e^{-ax} dF(x) + \int_{0}^{\infty} e^{ax} dF(x)$$

$$\leq \int_{-\infty}^{\infty} e^{-ax} dF(x) + \int_{-\infty}^{\infty} e^{ax} dF(x)$$

$$= m_X(-a) + m_X(a) < \infty$$

since $-a, a \in (-h, h)$.

Thus,

$$\frac{a^n}{n!} \int |x|^n \, dF(x) < \infty$$

and hence

$$\int |x|^n \, dF(x) < \infty$$

so that

$$E[X^n] = \int_\Omega X^n dP = \int_{-\infty}^{\infty} x^n dF_X(x)$$

$$= \int_{-\infty}^{\infty} x^n dF(x)$$

exists.

*Theorem** 4.3.30

If $m_X(\theta)$ exists for $|\theta| < h$, then

$$\left.\frac{d^n m_X(\theta)}{d\theta^n}\right|_{\theta=0} = \mu_n = E[X^n].$$

Example 4.3.31

Theorem (4.3.30) and *Example* (4.3.28) may be used to obtain the results of *Example* (4.3.15) parts ii) and iii) and *Example* (4.3.20) part ii).

Verification: i) Let X be as described in *Example* (4.3.15) part ii).

$$E[X] = \left.\frac{dm_X(\theta)}{d\theta}\right|_{\theta=0} = \left.\frac{d[pe^\theta + (1-p)]^n}{d\theta}\right|_{\theta=0}$$

$$= n[pe^\theta + (1-p)]^{n-1}\, pe^\theta\Big|_{\theta=0}$$

$$= n[p + (1-p)]^{n-1}p$$

$$= np$$

$$E[X^2] = \left.\frac{d^2m_X(\theta)}{d\theta^2}\right|_{\theta=0} = \{npe^\theta[(n-1)(pe^\theta + (1-p))^{n-2}]pe^\theta$$

$$+ [pe^\theta + (1-p)]^{n-1}npe^\theta\}\Big|_{\theta=0}$$

$$= np^2(n-1) + np$$
$$= n^2p^2 - np^2 + np;$$

ii) Let X be as described in *Example* (4.3.15) part iii).

$$E[X] = \left.\frac{de^{s(e^\theta-1)}}{d\theta}\right|_{\theta=0}$$

$$= e^{s(e^\theta-1)}se^\theta\Big|_{\theta=0}$$

$$= s$$

$$E[X^2] = \left.\frac{d^2m_X(\theta)}{d\theta^2}\right|_{\theta=0}$$

$$= \{se^\theta[e^{s(e^\theta-1)}se^\theta] + e^{s(e^\theta-1)}se^\theta\}\Big|_{\theta=0}$$

$$= s^2 + s;$$

iii) Let X be as described in *Example* (4.3.20) part ii).

$$E[X] = \frac{d(1-2\theta)^{-1}}{d\theta}\bigg|_{\theta = 0}$$

$$= -\frac{1}{(1-2\theta)^2}(-2)\bigg|_{\theta = 0}$$

$$= 2\,(1-2\theta)^{-2}\bigg|_{\theta = 0}$$

$$= 2$$

$$E[X^2] = \frac{d^2 m_X(\theta)}{d\theta^2}\bigg|_{\theta = 0}$$

$$= -4(1-2\theta)^{-3}(-2)\bigg|_{\theta = 0}$$

$$= 8.$$

We have seen above that $m_X(\theta)$, if it exists, generates the moments of a probability distribution. In addition, the *moment generating function* answers the important question: If we are given a set of moments, what is the *probability density function from which these moments came?* For the answer to this question, we state the following theorem.

*Theorem** 4.3.32

Let X and Y be two random variables with probability densities f(x) and g(y), respectively. If $m_X(\theta)$ and $m_Y(\theta)$ exist and are equal for all θ in the interval $|\theta| < \Theta$, $\Theta > 0$, then X and Y have the same probability distribution, $f(x) = g(y)$.

Example 4.3.33

Suppose that the random variable X has moment generating function

$$m_X(\theta) = \frac{\alpha}{\alpha - \theta}, \quad |\theta| < \alpha.$$

If the random variable Y has the following function for its probability density:

$$g(y) = \begin{cases} \alpha e^{-\alpha y} & , \quad y > 0,\ \alpha > 0 \\ 0 & , \quad \text{otherwise.} \end{cases}$$

Then, by *Theorem* (4.3.32) the probability density function of the random variable X is identical to that of the random variable Y. That is, $f(x) = g(y)$.

Verification: It is sufficient to show that $m_Y(\theta)$ exists and that

$$m_Y(\theta) = \frac{\alpha}{\alpha - \theta} \qquad |\theta| < \alpha.$$

By *Example* (4.2.13), Y has an exponential distribution with parameter $\frac{1}{\alpha}$.

By *Example* (4.3.28), $m_Y(\theta) = (1 - \frac{1}{\alpha}\theta)^{-1} = \dfrac{1}{\frac{\alpha-\theta}{\alpha}} = \dfrac{\alpha}{(\alpha-\theta)}$

for $|\theta| < \alpha$.

4.3 *EXERCISES*

1. Let X be a random variable such that $E[|X|^n] < \infty$ for $n \in N$. Let $\varepsilon > 0$. Prove that $P[|X| \geq \varepsilon] \leq E[|X|^n]/\varepsilon^n$

   Hint: Let $E = \{\omega : |X| \geq \varepsilon\}$. Calculate $E[|X|^n]$ by using *Definition* (4.3.1), *Theorem* (4.3.10).

2. (*Tchebychef's inequality*) Prove that if $E[X^2] < \infty$ then

$$P[|X-\mu| \geq \varepsilon] \leq \frac{\text{Var } X}{\varepsilon^2}, \quad \text{for } \varepsilon > 0.$$

3. Let X be a random variable such that Var (X) exists. Let a be a constant. Show that

$$\text{Var } a(X) = a^2 \text{ Var } (X)$$

4. Show that *Theorem* (4.3.10) is false if the condition

$$\sum_{i=1}^{\infty} \int_{E_i} |X| dP < \infty$$

is replaced by the condition

$$\sum_{i=1}^{\infty} \int_{E_i} X dP, \text{ exists.}$$

Hint: Consider $(\Omega, \mathcal{F}, P)$ where $\Omega = (0, 1]$ and P is the Lebesgue measure. Let

$E_i = (\frac{1}{i+1}, \frac{1}{i}]$ $i = 1, 2, 3, \ldots$ Define $X : \Omega \to R$ by $X(\omega) = (-1)^i (i+1)$ for $\omega \in E_i$.

Show that even though

$$\int_{E_i} X dP$$

exists for each i and

$$\sum_{i=1}^{\infty} \int X dP$$

exists, $\int X dP$ fails to exist by proving that

$$\int |X| dP = E[|X|]$$

does not exist.

5. Find the density function f for the discrete random variable X defined in *Exercise* (4.3.4) and show that even though $\sum_{\text{all } x} xf(x)$ in finite $E[X]$ does not exist.

6. Let X be a random variable of the continuous type with density f. Show that if $E[X]$ exists then

$$\int_{-\infty}^{\infty} xf(x)\,dx$$

is finite.

7. Prove that if $E[X^n]$ exists for n a natural number $E[X^r]$ exists for r a natural number, $r \leq n$. *Hint:* Note that $|X^r| \leq 1 + |X^n|$ and apply *Theorem* (4.3.3).

8. Let X be a random variable with uniform distribution over $[a, b]$. Use moment generating function techniques to find $E[X]$.

4.4 *SUMMARY*

The study of *random variables* was begun with emphasis being placed on the *theory* and *properties* of *one dimensional random variables*. The following was taken as the definition of the above:

Let $(\Omega, \mathcal{J}, P)$ be a probability space. Let $X : \Omega \to R$. X is said to be a one *dimensional random variable* if and only if X is *measurable*.

Sums and *products* of random variables were discussed and theorems relative to the composition of random variables with Borel measurable functions, continuous functions, and monotonic functions were obtained.

The *distribution function* for a random variable X was defined to be the function

$$F : \mathcal{R} \to \mathcal{R} \text{ such that}$$
$$F(x) = P\{\omega : X(\omega) \leq x\}.$$

F was shown to be a function which satisfied the following three conditions:

i) If $x_1 \leq x_2$, then $F(x_1) \leq F(x_2)$;

ii) F is continuous from the right at each point;

iii) $\lim_{x\to\infty} F(x) = 1$ and $\lim_{x\to-\infty} F(x) = 0$.

Two specific types of random variables were distinguished, namely, those which are *discrete* and those which are *continuous*. These were defined as follows:
Let $(\Omega, \mathcal{J}, P)$ be a probability space and let X be a random variable defined on Ω. X is said to be discrete if its range is countable. The non-negative function f defined on $\mathcal{R}$ by

$$f(x) = P[X = x]$$

is called the *density function* of X.
Let $(\Omega, \mathcal{J}, P)$ be a probability space and let X be a random variable defined on X. X is said to be of the *continuous type* if there exists a non-negative function f defined on $\mathcal{R}$, called the *density function* of X, such that

$$F(x) = P[X \leq x] = \int_{-\infty}^{x} f(t)dt$$

where the integral involved is the ordinary Riemann integral. The *point binomial*, *binomial* and *Poisson distributions* were considered as examples of discrete distributions while the *uniform*, *normal*, *Cauchy* and *exponential distributions* were presented as illustrations of random variables of the continuous type.

The general method of computing probabilities by means of the *Lebesgue integral* was considered. The implications of this procedure to the computation of probabilities in the case of discrete and continuous random variables were explored in detail.

The term *expectation* of X was defined in the following manner:

Let X be a random variable relative to $(\Omega, \mathcal{F}, P)$ integrable over Ω in the sense of Lebesgue. The *expectation* of X denoted $E[X]$, is defined by

$$E[X] = \int_{\Omega} X dP.$$

A series of theorems were presented which outlined the major properties of expectation and again the implications of these ideas in the case of discrete and continuous random variables were explored. The concepts of *ordinary moments*, *central moments* and *moment generating functions* which are defined in terms of expectation were introduced.

There were three primary objectives in this chapter:

i) *To study the mathematical properties of random*

variables and expectation;

ii) *To link the measure theoretic definition of the term random variable to those definitions commonly seen in elementary courses in probability and mathematical statistics, namely, the discrete and continuous cases;*

iii) *To link the definition of expectation in terms of the Lebesgue integral to those definitions presented in beginning courses for discrete and continuous random variables.*

IMPORTANT TERMS IN CHAPTER FOUR

random variable
distribution function
discrete random variable
density function
continuous random variable
point binomial distribution
binomial distribution
Poisson distribution
uniform distribution
normal distribution
Cauchy distribution
exponential distribution
expectation
proposition S *holds with probability one*
ordinary moments of X
central moments of X
variance of X
standard deviation of X
moment generating function for X

REFERENCES
AND
SUGGESTIONS FOR FURTHER READINGS

[1] Bartle, R. G., The Elements of Integration. New York: John Wiley and Sons, Inc., 1966.

[2] **Burrill, C. W., Measure, Integration and Probability. New York: McGraw-Hill Book Company, 1972.

[3] Gnedenko, B. V., The Theory of Probability. New York: Chelsea Publishing Company, 1966.

[4] **Kingman, J. F. C. and Taylor, S. J., Introduction to Measure and Probability. London: Cambridge University Press, 1966.

[5] Royden, H. L., Real Analysis, New York: The MacMillan Company, 1963.

[6] **Taylor, Angus E., General Theory of Functions and Integration. New York: Blaisdell Publishing Company, 1965.

[7] Tsokos, C. P., Probability Distributions: An Introduction to Probability Theory with Applications. Belmont, California: Wadsworth Publishing Company, Inc., 1972.

[8] **Tucker, H. G., A Graduate Course in Probability, New York: Academic Press, 1967.

[9] Widder, David V., Advanced Calculus. Englewood Cliffs, New Jersey: Prentice-Hall, Inc. 1961.

**These books are more advanced than the approach of the present text.

CHAPTER FIVE

MODES OF CONVERGENCE

5.0 *INTRODUCTION*

We shall discuss in this chapter various *modes of convergence* associated with sequences (X_n) of random variables relative to a probability space $(\Omega, \mathcal{F}, P)$. Since random variables are essentially just measurable functions relative to a measure space of measure one, most of the terminology employed has a direct measure theoretic analogue which we shall point out.

It is possible in certain pathological examples for a sequence of random variables to converge in some sense to a function defined on Ω which is not a random variable. In particular, this undesirable situation can arise if $(\Omega, \mathcal{F}, P)$ is not complete in the sense of *Definition* (3.3.1). This fact will be illustrated in *Example* (5.1.8). In the definitions which follow, when we say that a sequence (X_n) of random variables converges to X we shall assume that X is a random variable even if this fact is not stated explicitly.

5.1 *TYPES OF CONVERGENCE*

The relationships indicated in the following table will be pointed out in this chapter. They are summarized here for future easy reference.

Definition 5.1.1

Let $(\Omega, \mathcal{F}, P)$ be a probability space and let $A \in \mathcal{F}$. Let (X_n) be a sequence of random variables defined on Ω. The sequence (X_n) is said to *converge pointwise to the random variable* X on A, denoted $(X_n) \to X$, if given any $\varepsilon > 0$ and $\omega \in A$ there exists a natural number $N_{\varepsilon,\omega}$ such that for

$$n > N_{\varepsilon,\omega} ,$$

$$|X_n(\omega) - X(\omega)| < \varepsilon.$$

Note 5.1.2

If set A in the above definition is the sample space Ω, then we say simply that (X_n) converges to X everywhere or just that (X_n) converges to X. This definition is analogous to the concept of "pointwise" convergence of sequences of measurable functions in real analysis. The reason for the term pointwise is evident since the particular value of the natural number $N_{\varepsilon,\omega}$ required in the definition usually depends on ω as well as on ε as indicated by the subscripts. If the choice of $N_{\varepsilon,\omega}$ is actually independent of the choice of $\omega \in A$ then we say that (X_n) converges to X *uniformly* on A or Ω.

Theorem 5.1.3

If (X_n) converges to X uniformly on A, then (X_n) converges pointwise to X on A.

DICTIONARY OF PROBABILISTIC TERMINOLOGY

	Measure Theoretic Concepts		*Probabilistic Concepts*	
	Notation	*Terminology*	*Notation*	*Terminology*
1.	$(X_n) \to X$	the sequence (X_n) of measurable functions converges pointwise on Ω to the measurable function X.	$(X_n) \to X$	the sequence (X_n) of random variables converges everywhere to the random variable X.
2.	$(X_n) \to X$ uniformly	the sequence (X_n) of measurable functions converges uniformly on Ω to the measurable function X.	$(X_n) \to X$	the sequence (X_n) of random variables converges to the random variable X uniformly on Ω.
3.	$(X_n) \to X$ a.e.	the sequence (X_n) of measurable functions converges to the measurable function X almost everywhere.	$(X_n) \to X$ P.a.e.	the sequence (X_n) of random variables converges to the random variable X with probability one.

DICTIONARY OF PROBABILISTIC TERMINOLOGY

Measure Theoretic Concepts		*Probabilistic Concepts*	
Notation	*Terminology*	*Notation*	*Terminology*
4.	the sequence (X_n) of measurable functions converges to the measurable function X almost uniformly.		the sequence (X_n) of random variables converges to the random variable X almost uniformly.
5.	the sequence (X_n) of measurable functions converges in measure to the measurable function X.	$(X_n) \overset{P}{\to} X$	the sequence (X_n) of random variables converges in probability to the random variable X.

Examples 5.1.4

i) Let $(\Omega, \mathcal{F}, P)$ be a probability space. Define a sequence (X_n) of random variables by $X_n(\omega) = \frac{1}{n}$. $(X_n) \to 0$ uniformly on Ω.

ii) Let

$$(\Omega, \mathcal{F}, P) = (N, \rho_N, P)$$

where P is defined by

$$P\{\omega\} = \frac{1}{2^\omega} ;$$

$P[\phi] = 0$; and $P[E] = \sum_{\omega \varepsilon E} P\{\omega\}$ for $E \varepsilon \rho_N$, $E \neq \phi$.

Define a sequence (E_n) of elements of $\mathcal{F}$ by

$$E_n = \{\omega : \omega \leq n\}.$$

Define a sequence (X_n) of random variables by $X_n(\omega) = \chi_{E_n}(\omega)$. Then

$$(X_n) \to \chi_N$$

pointwise but the convergence is not uniform.

Proof: i) Choose $\varepsilon > 0$. Note that since $\lim_{n\to\infty} \frac{1}{n} = 0$, there exists a natural number N_ε such that for $n \geq N_\varepsilon$, $|\frac{1}{n} - 0| = |X_n(\omega) - 0| < \varepsilon$. Since the choice of N_ε is independent of ω, $(X_n) \to 0$ uniformly.

ii) Choose $\varepsilon > 0$ and let $\omega \varepsilon \Omega$. Choose $N_{\varepsilon,\omega} = \omega$. Note that for $n \geq N_{\varepsilon,\omega}$, $\omega \varepsilon E_n$ implying that $X_n(\omega) = \chi_{E_n}(\omega) = 1$. Thus for $n \geq N_{\varepsilon,\omega}$, $|X_n(\omega) - \chi_N(\omega)| = |1 - 1| = 0 < \varepsilon$ and $(X_n) \to \chi_N$ pointwise. The fact that the convergence is not uniform is evident.

Definition 5.1.5

Let (X_n) be a sequence of random variables defined relative to the probability space $(\Omega, \mathcal{F}, P)$. (X_n) is said to converge to the random variable X *with probability one* if there exists an element $A \varepsilon \mathcal{F}$ such that $P[A] = 0$ and (X_n) converges to X pointwise on the complement of A. We also say that (X_n) converges to X almost everywhere or P-*almost everywhere*. In this case we write

$$(X_n) \underset{P.a.e.}{\rightarrow} X \quad \text{or} \quad (X_n) \underset{a.e.}{\rightarrow} X.$$

This type of convergence is referred to simply as almost everywhere convergence in measure theoretic terms.

Note 5.1.6

The following statements are equivalent to *Definition* (5.1.5):

i) Let S be the statement "(X_n) converges to X pointwise." (X_n) converges to X with probability one if S is true with probability one;

ii) Let S be as in part i). (X_n) converges to X with probability one if

$$\{\omega : S \text{ fails}\} \varepsilon \mathcal{F} \quad \text{and} \quad P\{\omega : S \text{ fails}\} = 0;$$

iii) (X_n) converges to X with probability one if

$$P\{\omega : \lim_{n\to\infty} X_n(\omega) = X(\omega)\} = 1$$

and

$$\{\omega : \lim_{n\to\infty} X_n(\omega) = X(\omega)\} \in \mathcal{F}.$$

Theorem 5.1.7

If (X_n) converges to X everywhere then (X_n) converges to X with probability one.

Examples 5.1.8

i) Let $(\Omega, \mathcal{F}, P)$ be a probability space and let $F \in \mathcal{F}$ with $P[F] = 0$. Let $E \subseteq F$ but $E \notin \mathcal{F}$. That is, assume that $(\Omega, \mathcal{F}, P)$ is not complete in the sense of *Definition* (3.3.1). Define a sequence (X_n) of random variables by

$$(X_n)(\omega) \equiv 0. \quad (X_n) \underset{P.a.e.}{\to} \chi_E$$

but χ_E is not a random variable.

ii) Let $(\Omega, \mathcal{F}, P)$ be the probability space such that $\Omega = (0, 1)$; $\mathcal{F}$ = collection of Borel subsets of Ω; P = Borel measure. Define a sequence (X_n) of random variables by

$$\begin{cases} X_n(\omega) = 1 & \omega \notin Q \cap (0, 1) \\ X_n(\omega) = 0 & \omega \in Q \cap (0, 1) \end{cases}$$

where Q is the set of rational numbers. $(X_n) \underset{P.a.e.}{\to} 1$ but $(X_n) \not\to 1$ everywhere.

Verification: i) Note that since $X_n(\omega) \equiv 0$, $X_n(\omega) = \chi_E(\omega)$ on $\Omega - F \in \mathcal{F}$. Since $P[F] = 0$, we may apply *Definition* (5.1.5) to conclude that $(X_n) \underset{P.a.e.}{\to} \chi_E$.

Let $\alpha = \frac{1}{2}$. Note that $\chi_E^{-1}(\frac{1}{2}, \infty) = E \notin \mathcal{F}$ implying by *Theorem* (2.1.38) and *Definition* (4.1.1) that χ_E is not a random variable.

ii) Note that since Q is countable $Q \cap (0, 1)$ is countable by *Theorem* (1.1.13). By *Theorem* (2.1.32), $P[Q \cap (0, 1)] = 0$. Since $X_n(\omega) \equiv 1$ on the complement of $Q \cap (0, 1)$ we may apply *Definition* (5.1.5) to conclude that $(X_n) \underset{P.a.e.}{\to} 1$. Note that if $\omega \in Q \cap (0, 1)$, then $|X_n(\omega) - 1| = 1 \nless \varepsilon$ for $\varepsilon < 1$. Thus $(X_n) \not\to 1$ everywhere.

Definition 5.1.9

Let (X_n) be a sequence of random variables defined relative to the probability space $(\Omega, \mathcal{F}, P)$. (X_n) is said to converge *almost uniformly* to the random variable X if for each $\sigma > 0$ there exists an element $E_\sigma \in \mathcal{F}$ with $P[E_\sigma] < \sigma$ such that (X_n) converges uniformly to X on $\Omega - E_\sigma$. This type of convergence is referred to as almost uniform convergence in measure theoretic terms also.

Theorem 5.1.10

(X_n) converges to X almost uniformly if and only if (X_n) converges to X with probability one.

Proof: Let $(\sigma_k) = (\frac{1}{k})$. Assume that (X_n) converges to X almost uniformly. By *Definition* (5.1.9) for each $k \in N$

there exists an element $E_k \in \mathcal{S}$ such that $P[E_k] < \frac{1}{k}$ and such that (X_n) converges uniformly to X on $\Omega - E_k$. Consider any such sequence (E_k). Let $A = \bigcap_{k=1}^{\infty} E_k$. Note that since $\mathcal{S}$ is a σ algebra, $\bigcap_{k=1}^{\infty} E_k \in \mathcal{S}$. Note also that since $\bigcap_{k=1}^{\infty} E_k \subseteq E_k$ for each k, $P[\bigcap_{k=1}^{\infty} E_k] \leq P[E_k] < \frac{1}{k}$ by *Theorem* (3.1.5). This statement implies that $P[\bigcap_{k=1}^{\infty} E_k] \leq \varepsilon$ for any $\varepsilon > 0$ and hence that $P[\bigcap_{k=1}^{\infty} E_k] = P[A] = 0$. By *Theorem* (5.1.3) for each k (X_n) converges to X pointwise on $\Omega - E_k$ and hence also (X_n) converges to X pointwise on $\bigcup_{k=1}^{\infty} (\Omega - E_k)$. By *Theorem* (1.1.20) $\Omega - A = \Omega - \bigcap_{k=1}^{\infty} E_k = \bigcup_{k=1}^{\infty} (\Omega - E_k)$. Thus by *Definition* (5.1.5) we may conclude that (X_n) converges to X with probability one. The proof of the converse, called *Egoroff's Theorem* in measure theoretic terms, is given below. This proof entails the use of many of the results obtained in the previous material and should provide good exercise in their use.

Proof of Egoroff's Theorem

To reverse the argument, let $\Omega_0 = \{\omega \in \Omega : X_n(\omega) \not\to X(\omega)\}$. By *Note* (5.1.6), $\Omega_0 \in \mathcal{S}$ and $P[\Omega_0] = 0$. For each pair of natural numbers k and m let us define a set E_{km} by

$$E_{km} = \{\omega : |X_n(\omega) - X(\omega)| < \frac{1}{2^k} \text{ for } n \geq m\} \cap \Omega_0' .$$

Note that for each fixed value of k, we can express E_{km} by

$$E_{km} = \bigcap_{i=m}^{\infty} E_i$$

where for each i E_i is given by

$$E_i = \{\omega : |X_i(\omega) - X(\omega)| < \frac{1}{2^k}\} \cap \Omega_0' .$$

Written in this form it is evident that $E_{km} \in \mathcal{S}$ for each k and m. Note that $E_{km} \subseteq E_{k(m+1)}$ and that $E_{km} \subseteq \Omega - \Omega_0$. Thus

$$\bigcup_{m=1}^{\infty} E_{km} \in \mathcal{S}$$

and

$$\bigcup_{m=1}^{\infty} E_{km} \subseteq \Omega - \Omega_0 .$$

Let $\omega \in \Omega - \Omega_0$. Thus, by definition of Ω_0,

$$X_n(\omega) \to X(\omega) .$$

By *Definition* (5.1.1) given the positive number $\frac{1}{2^k}$ there exists a natural number N_k such that $n \geq N_k$ implies

$$|X_n(\omega) - X(\omega)| < \frac{1}{2^k} .$$

Hence by definition $\omega \in E_{kN_k}$ and also $\omega \in \bigcup_{m=1}^{\infty} E_{km}$. We thus have that

$$\Omega - \Omega_o \subseteq \bigcup_{m=1}^{\infty} E_{km}$$

and may use *Definition* (1.1.6) to conclude that

$$\bigcup_{m=1}^{\infty} E_{km} = \Omega - \Omega_o.$$

By *Theorem* (3.1.9)

$$\begin{aligned} P[\bigcup_{m=1}^{\infty} E_{km}] = P[\Omega - \Omega_o] &= P[\Omega] - P[\Omega_o] \\ &= 1 - 0 \\ &= 1. \end{aligned}$$

By *Theorem* (3.1.11)

$$P[\bigcup_{m=1}^{\infty} E_{km}] = \lim_{m\to\infty} P[E_{km}].$$

Combining these results we obtain that

$$\lim_{m\to\infty} P[E_{km}] = 1.$$

Consequently, given $\varepsilon > 0$, there exists a natural number M_k such that for $m \geq M_k$

$$\begin{aligned} |P[E_{km}] - 1| &= |P[E_{km}] - P[\Omega]| \\ &= P[\Omega] - P[E_{km}] \\ &= P[\Omega - E_{km}] < \frac{\varepsilon}{2^k}. \end{aligned}$$

In particular for each k,

$$P[\Omega - E_{kM_k}] < \frac{\varepsilon}{2^k}.$$

Let

$$E = \Omega - \bigcap_{k=1}^{\infty} E_{kM_k} = \bigcup_{k=1}^{\infty} [\Omega - E_{kM_k}],$$

by *Theorem* (1.1.20).

Thus

$$\begin{aligned} P[E] &= P[\bigcup_{k=1}^{\infty} [\Omega - E_{kM_k}]] \\ &\leq \sum_{k=1}^{\infty} P[\Omega - E_{kM_k}] \quad \text{by } \textit{Theorem } (3.1.14) \\ &< \sum_{k=1}^{\infty} \frac{\varepsilon}{2^k} \\ &= \varepsilon \sum_{k=1}^{\infty} \frac{1}{2^k} \\ &= \varepsilon. \end{aligned}$$

We need only show that (X_n) converges to X uniformly on $\Omega - E$ and the proof will be complete. Note that

$$\Omega - E = \Omega - [\Omega - \bigcap_{k=1}^{\infty} E_{kM_k}] = \bigcap_{k=1}^{\infty} E_{kM_k},$$ by *Theorem* (1.1.17). Thus for $\omega \varepsilon \Omega - E$ and $n \geq M_k$ we have by definition of E_{kM_k} that

$$|X_n(\omega) - X(\omega)| < \frac{1}{2^k}$$

for any $k = 1, 2, 3, \ldots$, which implies that the convergence is uniform on $\Omega - E$.

Definition 5.1.11

Let (X_n) be a sequence of random variables defined relative to the probability space $(\Omega, \mathcal{F}, P)$. (X_n) is said *to converge in probability to the random variable* X if

$$\lim_{n\to\infty} P\{\omega : |X_n(\omega) - X(\omega)| \geq \alpha\} = 0$$

for each $\alpha > 0$. We shall denote this type of convergence by

$$(X_n) \overset{P}{\to} X.$$

This type of convergence is referred to as convergence in measure in the measure theoretic context.

Theorem 5.1.12

If (X_n) converges to X almost uniformly then (X_n) converges to X in probability.

Proof: Assume that (X_n) converges to X almost uniformly. Assume that $\alpha > 0$ and $\varepsilon > 0$. By *Definition* (5.1.9) there exists an element $E_\varepsilon \in \mathcal{F}$ such that $P[E_\varepsilon] < \varepsilon$ and (X_n) converges to X uniformly on $\Omega - E_\varepsilon$. Note that by *Note* (5.1.2) there exists a natural number N_α such that for $n \geq N_\alpha$ we have $|X_n(\omega) - X(\omega)| < \alpha$ for all $\omega \in \Omega - E_\varepsilon$. Thus for $n \geq N_\alpha$,

$$\{\omega : |X_n(\omega) - X(\omega)| \geq \alpha\} \subseteq \Omega - [\Omega - E_\varepsilon] = E_\varepsilon.$$

This is precisely what is required in order to show that $\lim_{n\to\infty} P\{\omega : |X_n(\omega) - X(\omega)| \geq \alpha\} = 0$.

Thus by *Definition* (5.1.11) (X_n) converges to X in probability.

Definition 5.1.13

Let (X_n) be a sequence of random variables relative to the probability space $(\Omega, \mathcal{F}, P)$ such that $E[X_n^2] < \infty$ for each n. (X_n) is said to *converge in mean square* to a random variable X such that $E[X^2] < \infty$ if

$$\lim_{n\to\infty} E[|X_n - X|^2] = 0.$$

We shall write

$$(X_n) \underset{\text{m.s.}}{\to} X .$$

This type of convergence would be termed convergence in mean of order two in a general measure space.

Theorem 5.1.14

If (X_n) converges to X in mean square then (X_n) converges to X in probability.

Proof: Assume that (X_n) converges to X in mean square. Choose $\alpha > 0$. For each n let $E_n(\alpha)$ be defined by

$$E_n(\alpha) = \{\omega : |X_n(\omega) - X(\omega)| \geq \alpha\} = \{\omega : |X_n(\omega) - X(\omega)|^2 \geq \alpha^2\}.$$

Thus $E[|X_n - X|^2] = \int |X_n - X|^2 \, dP$ by *Definition* (4.3.1)

$$\geq \int_{E_n(\alpha)} |X_n - X|^2 \, dP$$ by *Theorem* (2.2.21)

$$\geq \int_{E_n(\alpha)} \alpha^2 \, dP$$ by *Theorem* (2.2.20) and *Definition* (2.2.18)

$$= \alpha^2 P[E_n(\alpha)].$$

Since $\lim_{n\to\infty} E[|X_n - X|^2] = 0$, given $\varepsilon > 0$ there exists a natural number N_ε such that $n > N_\varepsilon$ implies $E[|X_n - X|^2] < \alpha \; \varepsilon$. Thus also for $n > N_\varepsilon$ we have

$$\alpha^2 P[E_n(\alpha)] \leq E[|X_n - X|^2] < \varepsilon \, \alpha^2$$

or

$$P[E_n(\alpha)] < \varepsilon.$$ Hence

$$\lim_{n\to\infty} P\{\omega : |X_n(\omega) - X(\omega)| \geq \alpha\} = 0$$

and the proof is complete.

Examples 5.1.15

i) Let $(\Omega, \mathcal{F}, P)$ be the probability space such that $\Omega = (0, 1)$; $\mathcal{F}$ = collection of Borel subsets of Ω; P = Borel measure. Consider the intervals

$$(0, 1), \; (0, \tfrac{1}{2}), \; (\tfrac{1}{2}, 1), \; (0, \tfrac{1}{3}), \; (\tfrac{1}{3}, \tfrac{2}{3}), \; (\tfrac{2}{3}, 1),$$

$(0, \frac{1}{4})$, $(\frac{1}{4}, \frac{1}{2})$, $(\frac{1}{2}, \frac{3}{4})$, $(\frac{3}{4}, 1)$, $(0, \frac{1}{5})$,

$(\frac{1}{5}, \frac{2}{5})$, ...

Define a sequence (X_n) of random variables on Ω by $X_n(\omega) = \chi_{E_n}(\omega)$ where E_n is the nth interval of the above list. (X_n) converges to 0 in mean square but (X_n) does not converge to 0 pointwise at <u>any</u> point of $(0, 1)$. Thus also (X_n) does not converge to 0 with probability one or almost uniformly.

ii) Let $(\Omega, \mathcal{S}, P)$ be as described in part i). Define a sequence of events (E_n) by $E_n = (\frac{1}{n+1}, \frac{1}{n})$. Define a sequence of random variables on Ω by $X_n(\omega) = (n^2 + n)\chi_{E_n}(\omega)$. (X_n) converges to 0 everywhere on Ω but (X_n) does not converge to 0 in mean square.

iii) Let $(\Omega, \mathcal{S}, P)$ be the probability space such that $\Omega = (0, 1]$; $\mathcal{S}$ = collection of Borel subsets of Ω; P = Borel measure. Define a sequence of events (E_n) by

$$E_n = (\frac{p}{2q}, \frac{p+1}{2q})$$

where p and q are the unique integers such that

$$p + 2^q = n \quad \text{and} \quad 0 \leq p < 2^q.$$

Define a sequence (X_n) of random variables on Ω by $X_n(\omega) = \chi_{E_n}(\omega)$. (X_n) converges in probability to 0 but (X_n) does not converge to 0 pointwise for any point of (0, 1]. Hence, also (X_n) does not converge to 0 with probability one or almost uniformly.

Verification: i) Let (X_n) and E_n be as described in the example. Note that since we are considering Borel measure, $P[E_n]$ is simply the length of the interval in question and due to the method of construction $\lim_{n\to\infty} P[E_n] = 0$. Now

$$E[|X_n - X|^2] = \int |X_n - X|^2 \, dP$$

$$= \int X_n^2 \, dP$$

$$= \int_{E_n} X_n^2 \, dP$$

$$= \int_{E_n} dP$$

$$= P[E_n].$$

Thus $\lim_{n\to\infty} E[|X_n - X|^2] = \lim_{n\to\infty} P[E_n] = 0$ and (X_n) converges to 0 in mean square. To see that (X_n) converges to 0 nowhere on (0, 1), simply note that for any $\omega \varepsilon (0, 1)$ the sequence $(X_n(\omega))$ will have one subsequence which consists entirely of zeroes and another which consists entirely of ones and hence it cannot converge.

ii) Let $\omega \in \Omega$. There exists a natural number N_ω such that $n \geq N_\omega$ implies $\frac{1}{n} < \omega$. Thus for $n \geq N_\omega$, $\omega \notin E_n$ and thus $X_n(\omega) = (n^2 + n)\chi_{E_n}(\omega) = 0$ implying that (X_n) converges to 0 everywhere. To see that (X_n) does not converge to 0 in mean square, note that since we are dealing with Borel measure,

$P[E_n] = \frac{1}{n} - \frac{1}{n+1} = \frac{1}{n^2+n}$. Now

$$\begin{aligned} E[|X_n - X|^2] &= E[|X_n|^2] \\ &= \int X_n^2 \, dP \\ &= \int_{E_n} (n^2 + n) \, dP \\ &= (n^2 + n) \int_{E_n} dP \\ &= (n^2 + n) P[E_n] \\ &= (n^2 + n) \frac{1}{n^2+n} \\ &= 1. \end{aligned}$$

Thus $\lim_{n\to\infty} E[|X_n - X|^2] = \lim_{n\to\infty} 1 = 1 \neq 0$ and (X_n) does NOT converge to 0 in mean square.

iii) Note first the pattern created by considering the sequence of events (E_n). Namely $(0, 1]$, $(0, 1/2]$, $(1/2, 1]$, $(0, 1/4]$, $(1/4, 2/4]$, $(2/4, 3/4]$, $(3/4, 1]$, $(0, 1/8]$, $(1/8, 2/8]$, $(2/8, 3/8]$, $(3/8, 4/8]$, $(4/8, 5/8]$, ... It is evident that for any $\omega \in \Omega$, the sequence $(X_n(\omega))$ will have one subsequence which consists entirely of ones and another which will consist entirely of zeroes and hence $(X_n(\omega))$ cannot converge. Choose $\alpha > 0$. If $\alpha > 1$, then

$$\{\omega : |X_n(\omega) - X(\omega)| \geq \alpha\} = \{\omega : X_n(\omega) \geq \alpha\} = \phi$$

and hence

$$\lim_{n\to\infty} P\{\omega : |X_n(\omega) - X(\omega)| \geq \alpha\} = \lim_{n\to\infty} 0 = 0.$$

If $\alpha \leq 1$ then

$$\{\omega : |X_n(\omega) - X(\omega)| \geq \alpha\} = \{\omega : X_n(\omega) \geq \alpha\} = E_n.$$

Hence $\lim_{n\to\infty} P\{\omega : |X_n(\omega) - X(\omega)| \geq \alpha\} = \lim_{n\to\infty} P[E_n]$.

However since we are dealing with Borel measure,

$$P[E_n] = \frac{p+1}{2^q} - \frac{p}{2^q} = \frac{1}{2^q} \quad \text{if}$$

$2^q \leq n < 2^{q+1}$. Expanding this sequence of real numbers for a few terms we obtain that $(P[E_n])$ is the sequence 1, 1/2, 1/2, 1/4, 1/4, 1/4, 1/4, 1/8, 1/8, 1/8, 1/8, 1/8, 1/8, 1/8, 1/8, 1/16, 1/16, . . . which obviously converges to 0. Hence (X_n) converges to 0 in probability.

5.2 *SUMMARY OF RELATIONSHIPS AMONG VARIOUS MODES OF CONVERGENCE*

In the charts below a solid arrow represents implications and a slashed arrow implications which will not hold.

The numbers given reference either a proof of implication or a counterexample. We shall employ the following notation:

c.e. (*convergence everywhere*)

c.u. (*convergence uniformly on* Ω)

c.a.e. (*convergence almost everywhere or with probability one*)

c.a.u. (*convergence almost uniformly*)

c.m.s. (*convergence in mean square*)

c.p. (*convergence in probability*)

$$\text{c.u.}\ \begin{array}{c}(5.1.3)\\ \longrightarrow\\ \not\longleftarrow\\ (5.1.4,\text{ii})\end{array}\ \text{c.e.}\ \begin{array}{c}(5.1.7)\\ \longrightarrow\\ \not\longleftarrow\\ (5.1.8,\text{ii})\end{array}\ \text{c.a.e.}$$

$$\text{c.e.}\ \begin{array}{c}(5.1.16,\text{ii})\\ \not\longrightarrow\\ \not\longleftarrow\\ (5.1.16,\text{i})\end{array}\ \text{c.m.s.}$$

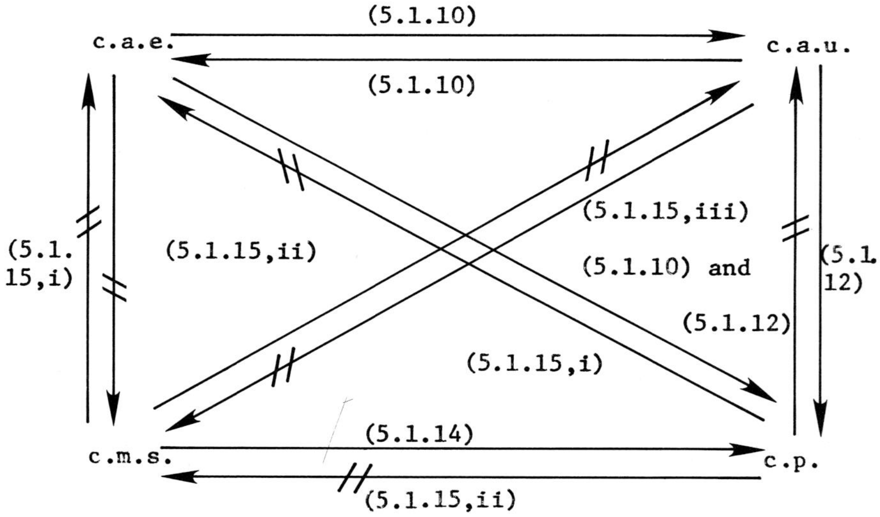

5.2 *EXERCISES*

1. Let (X_n) be a sequence of random variables defined relative to the probability space $(\Omega, \mathcal{F}, P)$ which converges to the random variable X with probability one. If $g : R \to R$ is continuous then the sequence $(g(X_n))$ converges to $g(X)$ with probability one.

2. The restriction that g be continuous is necessary in *Exercise* (5.2.1).

   Hint: Let x_o be a point of discontinuity of g. Let

$$D_n = \{x : |x - x_o| < \frac{1}{n}\}.$$

   Define a sequence (x_n) of real numbers where

$$x_n \in D_n$$

   and

$$|g(x_n) - g(x_o)| \geq \varepsilon^*.$$

   Define a sequence of random variables (X_n) and a random variable X by

$$X_n(\omega) \equiv x_n \quad \text{and} \quad X(\omega) \equiv x_o.$$

   Show that

$$(X_n) \underset{P.a.e.}{\to} X$$

   but

$$(g(X_n)) \underset{P.a.e.}{\not\to} g(X).$$

3. Let (X_n) be a sequence of random variables relative to the probability space $(\Omega, \mathcal{F}, P)$ such that $E[X_n^2] < \infty$ for each n. Show that if

$$(X_n) \underset{m.s.}{\to} X$$

then

$$(X_n) \underset{P}{\to} X$$

by use of *Exercise* (4.3.1).

5.3 SUMMARY

Six types or *modes of convergence* of sequences of random variables were considered. These were

i) *convergence everywhere;*
ii) *convergence uniformly on* Ω *;*
iii) *convergence with probability one;*
iv) *convergence almost uniformly;*
v) *convergence in mean square;*
vi) *convergence in probability.*

A detailed discussion of the relationships that exist among the above modes of convergence was presented.

IMPORTANT TERMS IN CHAPTER FIVE

pointwise convergence
convergence everywhere
uniform convergence on Ω
convergence with probability one
convergence almost uniformly
convergence in probability
convergence in mean square

REFERENCES AND SUGGESTIONS FOR FURTHER READINGS

[1] Bartle, R. G., The Elements of Integration. New York: John Wiley and Sons, Inc., 1966.

[2] **Burrill, C. W., Measure, Integration and Probability. New York: McGraw-Hill Book Company, 1972.

**This book is more advanced than the approach of the present text.

CHAPTER SIX

n-DIMENSIONAL RANDOM VARIABLES AND INDEPENDENCE

6.0 *INTRODUCTION*

In this chapter we shall introduce briefly the concept of n-*dimensional random variables*. Most of the terminology employed is a natural extension of that encountered in Chapter Four in the one-dimensional case. We shall not go into great detail concerning this topic as our primary concern is to define the idea of *independent random variables* so that we may in Chapter Seven consider many of the so called *limit theorems*. The reader who is interested in pursuing the study of n-dimensional random variables is referred in particular to Tucker, [3] and Gnedenko [2]. It would be helpful at this point to review *Definition* (2.1.45) to *Theorem* (2.1.51).

6.1 n-*DIMENSIONAL RANDOM VARIABLES*

The random variables discussed thus far have all been "one dimensional." We now extend this idea to consider n-dimensional random variables. The reader is referred to Chapter One, section two for a review of Euclidean n-space.

DICTIONARY OF PROBABILISTIC TERMINOLOGY

Measure Theoretic Concepts		*Probabilistic Concepts*	
Notation	*Terminology*	*Notation*	*Terminology*
f	$f = (f_1, f_2, \ldots, f_n)$ is a measurable function from the measure space $(X, \mathcal{C}, \lambda)$ into R^n.	$X = X_1, X_2, X_3, \ldots, X_n)$	X is an n-dimensional random variable or an n-dimensional random vector.

Definition 6.1.1

Let $(\Omega, \mathcal{F}, P)$ be a probability space. Let $X : \Omega \to R^n$. X is said to be an n-*dimensional random variable* if and only if X is measurable.

Note 6.1.2

n-dimensional random variables are often referred to as n-dimensional random vectors. In this case we use the usual vector notation $X = (X_1, X_2, X_3, \ldots, X_n)$ where for each $i = 1, 2, \ldots n$, X_i is a function from Ω into R and

$$X(\omega) = (X_1(\omega), X_2(\omega), \ldots, X_n(\omega)).$$

Theorem 6.1.3

Let $(\Omega, \mathcal{F}, P)$ be a probability space and $X = (X_1, X_2, \ldots, X_n)$ a map from $\Omega \to R^n$. X is an n-dimensional random variable if and only if for each $i = 1, 2, \ldots n$ X_i is a one-dimensional random variable.

The following theorem provides a useful extension of (4.1.16) to the n-dimensional case.

Theorem 6.1.4

Let $f : R^m \to R^n$ be Borel measurable. If X is an m-dimensional random variable, then $f \circ X$ or $f(X)$ is an n-dimensional random variable.

Proof: Let $f : R^m \to R^n$ be Borel measurable. Let X be an m-dimensional random variable. Let E be a Borel set in R^n. Note that
$\{\omega : (f \circ X)(\omega) \in E\} = \{\omega : X(\omega) \in f^{-1}(E)\} = X^{-1}(f^{-1}(E))$.

Since f is Borel measurable, $f^{-1}(E)$ is a Borel set in

$\mathcal{R}^m$ and since X is an m-dimensional random variable, $X^{-1}(f^{-1}(E)) \in \mathcal{S}$ implying that $f \circ X$ is an n-dimensional random variable.

Definition 6.1.5

Let $(\Omega, \mathcal{S}, P)$ be a probability space and $X = (X_1, X_2, \ldots, X_n)$ an n-dimensional random variable defined on Ω. Let $H : \mathcal{R}^n \to \mathcal{R}$ be defined by

$$H(x_1, x_2, x_3, \ldots, x_n) = P\{\omega : X_1(\omega) \leq x_1, X_2(\omega) \leq x_2, \ldots, X_n(\omega) \leq x_n\}.$$

H is called the *cumulative distribution function, the joint distribution function* or simply the *distribution function* of X.

Note 6.1.6

We shall employ a notational convention analogous to that of *Note* (4.2.2). Namely we shall write

$$P\{\omega : X_1(\omega) \leq x_1, X_2(\omega) \leq x_2, \ldots, X_n(\omega) \leq x_n\} =$$

$$P[X_1 \leq x_1, X_2 \leq x_2, \ldots, X_n \leq x_n].$$

Similarly we shall write such expressions as

$$P[X_1 = x_1, X_2 = x_2, \ldots, X_n = x_n],$$

$$P[X_1 < x_1, X_2 = x_2, \ldots, X_n \neq x_n]$$

and others.

Just as in the one dimensional case it will be useful to distinguish in particular two special types of n-dimensional random variables. Namely we shall define what we mean by n-dimensional discrete random variables and n-dimensional continuous random variables.

Definition 6.1.7

Let $(\Omega, \mathcal{S}, P)$ be a probability space and let X be an n-dimensional random variable defined on Ω. X is said to be *discrete* if its range is a countable subset of $\mathcal{R}^n$.

Definition 6.1.8

Let $X = (X_1, X_2, \ldots, X_n)$ be an n-dimensional discrete random variable. The non-negative function f defined on $\mathcal{R}^n$ by

$$f(x_1, x_2, \ldots, x_n) = P[X_1 = x_1, X_2 = x_2, \ldots, X_n = x_n]$$

is called the *probability density function, joint density function* or *simply the density function* of X.

Example 6.1.9

i) Consider an experiment which consists of tossing a fair coin three times. Denote a possible outcome of this experiment by an ordered triple of the form (H, T, H) where of course H represents heads and T tails. Let

$$\Omega = \{\omega_1, \omega_2, \omega_3, \omega_4, \omega_5, \omega_6, \omega_7, \omega_8\}$$

where

$\omega_1 = (H, H, H)$ $\qquad$ $\omega_5 = (T, T, H)$

$\omega_2 = (H, H, T)$ $\qquad$ $\omega_6 = (T, H, T)$

$\omega_3 = (H, T, H)$ $\qquad$ $\omega_7 = (H, T, T)$

$\omega_4 = (T, H, H)$ $\qquad$ $\omega_8 = (T, T, T)$.

Let $\mathcal{S} = \rho_\Omega$. Define a set function P on $\mathcal{S}$ by $P\{\omega_i\} = 1/8$, $i = 1, 2, \ldots 8$; for $E \in \mathcal{S}$, $E \neq \phi$

$$P[E] = \sum_{\omega \in E} P\{\omega\}; \quad P[\phi] = 0.$$

Define maps X_1 and X_2 from Ω into $\mathcal{R}$ by $X_1(\omega) =$ number of heads represented by ω, $X_2(\omega) =$ number of tails represented by ω. $(\Omega, \mathcal{S}, P)$ is a probability space and $X = (X_1, X_2)$ is a two dimensional discrete random variable. Furthermore the joint density f of (X_1, X_2) is given by

$$f(3, 0) = f(0, 3) = 1/8$$
$$f(2, 1) = f(1, 2) = 3/8$$
$$f(x_1, x_2) = 0 \quad \text{elsewhere.}$$

The joint distribution H of (X_1, X_2) is given by

I	$H(x_1, x_2) = 0$	$x_1 + x_2 < 3$
II	$H(x_1, x_2) = 0$	$x_1 + x_2 \geq 3$ and $x_1 < 0$
III	$H(x_1, x_2) = 0$	$x_1 + x_2 \geq 3$ and $x_2 < 0$
IV	$H(x_1, x_2) = 1$	$x_1 \geq 3$ and $x_2 \geq 3$
V	$H(x_1, x_2) = 1/8$	$0 \leq x_1 < 1$ and $3 \leq x_2$

VI	$H(x_1, x_2) = 0$	$0 \le x_1 < 1$, $x_1 + x_2 \ge 3$ and $3 > x_2$
VII	$H(x_1, x_2) = 0$	$1 \le x_1 < 2$ and $x_1 + x_2 \ge 3$ and $2 > x_2$
VIII	$H(x_1, x_2) = 0$	$2 \le x_1 < 3$, $x_1 + x_2 \ge 3$ and $1 > x_2$
IX	$H(x_1, x_2) = 3/8$	$1 \le x_1 < 2$, $2 \le x_2 < 3$
X	$H(x_1, x_2) = 1/2$	$1 \le x_1 < 2$, $3 \le x_2$
XI	$H(x_1, x_2) = 3/8$	$2 \le x_1 < 3$, $1 \le x_2 < 2$
XII	$H(x_1, x_2) = 6/8$	$2 \le x_1 < 3$, $2 \le x_2 < 3$
XIII	$H(x_1, x_2) = 7/8$	$2 \le x_1 < 3$, $3 \le x_2$
XIV	$H(x_1, x_2) = 1/8$	$3 \le x_1$, $0 \le x_2 < 1$
XV	$H(x_1, x_2) = 1/2$	$3 \le x_1$, $1 \le x_2 < 2$
XVI	$H(x_1, x_2) = 7/8$	$3 \le x_1$, $2 \le x_2 < 3$

Hint: To verify that the distribution function H is defined as above, break R^2 up into the sixteen regions indicated in the following diagram which correspond to the regions stated above.

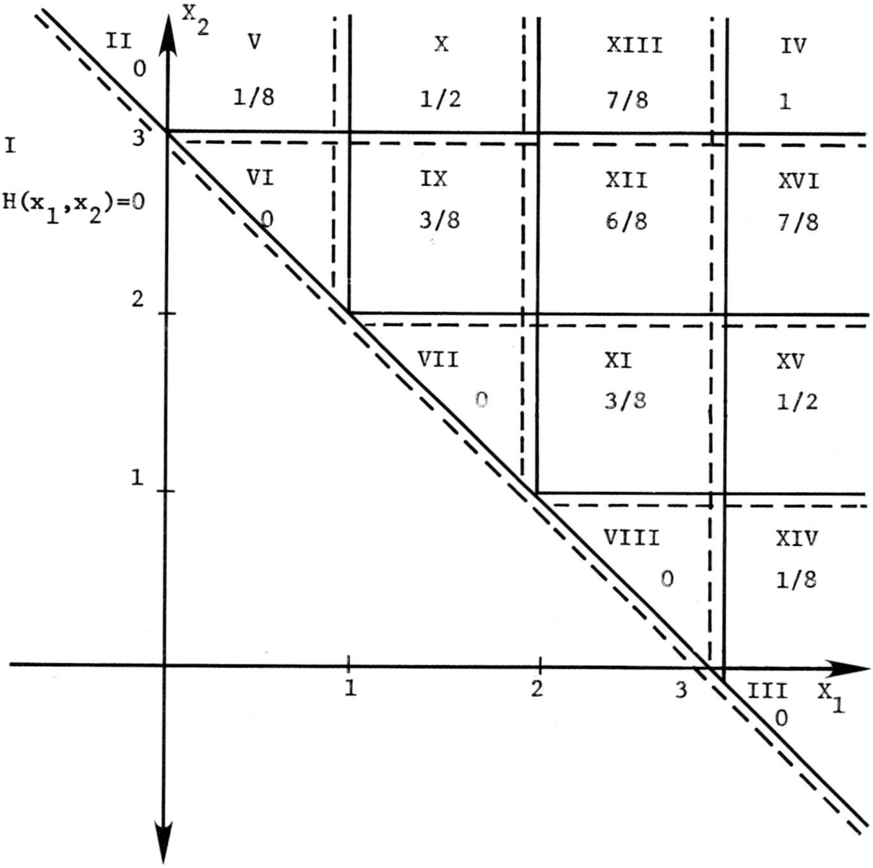

X_2
II
0
V
1/8
X
1/2
XIII
7/8
IV
1
I
3
$H(x_1,x_2)=0$
VI
0
IX
3/8
XII
6/8
XVI
7/8
2
VII
0
XI
3/8
XV
1/2
1
VIII
0
XIV
1/8
1
2
3
III
0
X_1

ii) Consider an experiment which consists of tossing a fair die twice and observing the outcome of each toss. Let

$$\Omega = \{(i, j) : 1 \leq i \leq 6, 1 \leq j \leq 6\}$$

$$\mathcal{S} = \rho_{\Omega}.$$

Define a set function P on $\mathcal{S}$ by

$$P\{\omega\} = 1/36 \quad \text{for each } \omega \in \Omega; \text{ for } E \in \mathcal{S}, E \neq \phi$$

$$P[E] = \sum_{\omega \in E} P\{\omega\}; \quad P[\phi] = 0.$$

Define maps X_1 and X_2 from Ω into $\mathcal{R}$ by

$$X_1(\omega) = |i - j|$$

$$X_2(\omega) = i + j.$$

$(\Omega, \mathcal{S}, P)$ is a probability space and $X = (X_1, X_2)$ is a two dimensional discrete random variable. Let $\mathcal{R}_X$ denote the range of X. $\mathcal{R}_X$ contains twenty-one elements. Furthermore, the joint density f of (X_1, X_2) is given by

$$f(x_1, x_2) = 1/36, \quad (x_1, x_2) \in \mathcal{R}_X, x_1 = 0$$

$$f(x_1, x_2) = 1/18, \quad (x_1, x_2) \in \mathcal{R}_X, x_1 > 0$$

$$f(x_1, x_2) = 0, \quad \text{elsewhere.}$$

iii) As was the case when dealing with one dimensional random variables, n-dimensional random variables are often defined by simply stating the joint density f. One of the most widely known n-dimensional discrete distributions is the so-called multinomial distribution defined below. "The k dimensional random variable $(X_1, X_2, \ldots, X_k)$ is said to have a multinomial distribution with parameters $n, p_1, p_2, \ldots, p_k$ if its joint density f is given by

$$f(x_1, x_2, \ldots, x_k) = \begin{cases} \dfrac{n!}{x_1!x_2!\ldots x_k!} p_1^{x_1} p_2^{x_2} \ldots p_k^{x_k}, & x_i = 0, 1, 2, \ldots n \\ & 0 \le p_i \le 1 \\ & i = 1, 2, \ldots, k \\ 0, & \text{elsewhere} \end{cases}$$

where $\sum_{i=1}^{k} p_i = 1$ and $\sum_{i=1}^{k} x_i = n$."

Verification: ii) To verify the assertions made in part ii) consider the following table:

ω	$X(\omega)$	R_X	Number of elements of Ω mapped by X into each value of R_X
(1, 1)	(0, 2)	(0, 2)	1
(1, 2)	(1, 3)	(1, 3)	2
(1, 3)	(2, 4)	(2, 4)	2
(1, 4)	(3, 5)	(3, 5)	2
(1, 5)	(4, 6)	(4, 6)	2
(1, 6)	(5, 7)	(5, 7)	2
(2, 1)	(1, 3)	(0, 4)	1

ω	$X(\omega)$	R_X	Number of elements of Ω mapped by X into each value of R_X
(2, 2)	(0, 4)	(1, 5)	2
(2, 3)	(1, 5)	(2, 6)	2
(2, 4)	(2, 6)	(3, 7)	2
(2, 5)	(3, 7)	(4, 8)	2
(2, 6)	(4, 8)	(0, 6)	1
(3, 1)	(2, 4)	(1, 7)	2
(3, 2)	(1, 5)	(2, 8)	2
(3, 3)	(0, 6)	(3, 9)	2
(3, 4)	(1, 7)	(0, 8)	1
(3, 5)	(2, 8)	(1, 9)	2
(3, 6)	(3, 9)	(2, 10)	2
(4, 1)	(3, 5)	(0, 10)	1
(4, 2)	(2, 6)	(1, 11)	2
(4, 3)	(1, 7)	(0, 12)	1
(4, 4)	(0, 8)		
(4, 5)	(1, 9)		
(4, 6)	(2, 10)		
(5, 1)	(4, 6)		
(5, 2)	(3, 7)		
(5, 3)	(2, 8)		
(5, 4)	(1, 9)		
(5, 5)	(0, 10)		
(5, 6)	(1, 11)		
(6, 1)	(5, 7)		
(6, 2)	(4, 8)		
(6, 3)	(3, 9)		
(6, 4)	(2, 10)		
(6, 5)	(1, 11)		
(6, 6)	(0, 12)		

Definition 6.1.10

Let $(\Omega, \mathcal{F}, P)$ be a probability space and let $X = (X_1, X_2, \ldots, X_n)$ be an n-dimensional random variable defined on Ω. X *is said to be of the continuous type* if there exists a non-negative function f defined on $\mathcal{R}^n$, called the joint density (or density) of X, such that

$$H(x_1, x_2, x_3, \ldots, x_n) = \int_{-\infty}^{x_1} \int_{-\infty}^{x_2} \cdots \int_{-\infty}^{x_n} f(t_1, t_2, \ldots, t_n)\, dt_1 dt_2 \ldots dt_n$$

where the integral involved is the ordinary Riemann integral and H is the joint distribution function of X.

Example 6.1.11

A two-dimensional random variable (X_1, X_2) is said to have a *bivariate normal distribution* with parameters s_1, s_2, r, a_1 and a_2 if its joint density f is given by

$$f(x_1, x_2) = \frac{1}{2\pi s_1 s_2 \sqrt{1-r^2}}\, e^{-\frac{1}{2(1-r^2)}\left[\left(\frac{x_1-a_1}{s_1}\right)^2 - \frac{2r(x_1-a_1)(x_2-a_2)}{s_1 s_2} + \left(\frac{x_2-a_2}{s_2}\right)^2\right]}$$

where $a_1, a_2, \in \mathcal{R}$; $s_1, s_2 > 0$; $-1 < r < 1$.

The above distribution is one of the most widely used in practice.

6.2 *INDEPENDENCE*

In this section we shall consider the topic of independence of random variables. As we shall see this idea is closely tied in with the concept of independent events presented in Chapter Three. Independence is an interesting concept in its own right and will play a key role in our discussion of some limit theorems to be presented in Chapter Seven. Proofs for starred theorems may be found in Tucker, [3] or Burrill [1].

Definition 6.2.1

Let $K = \{X_\gamma : \gamma \in \Gamma\}$ be a collection of one dimensional random variables defined relative to the probability space $(\Omega, \mathcal{F}, P)$. These random variables are said to be *stochastically independent* if for every natural number n and every n distinct elements $\gamma_1, \gamma_2, \ldots, \gamma_n \in \Gamma$ and any collection $\{B_1, B_2, \ldots, B_n\}$ of Borel subsets of $\mathcal{R}$ the events

$$\{X_{\gamma_1}^{-1}(B_1), X_{\gamma_2}^{-1}(B_2), \ldots, X_{\gamma_n}^{-1}(B_n)\}$$

are mutually independent. That is,

$$P[\bigcap_{i=1}^{n} X_{\gamma}^{-1}(B_i)] = \prod_{i=1}^{n} P[X_{\gamma}^{-1}(B_i)].$$

Note 6.2.2

We shall often refer to random variables which are stochastically independent as being simply "*independent*." Unless otherwise indicated the random variables referred to in this section are understood to be one dimensional.

When we say that $X_1, X_2, \ldots, X_n$ are independent we of course mean that $\{X_i : i = 1, 2, 3, \ldots, n\}$ satisfies *Definition* (6.2.1).

Theorem 6.2.3

Let $X_1, X_2, \ldots, X_n$ be independent random variables. For each i, let $f_i : \mathcal{R} \to \mathcal{R}$ be Borel measurable. (See *Note* (2.1.40)). Then the random variables $f_1(X_1), f_2(X_2), \ldots, f_n(X_n)$ are independent.

Proof: Note that by *Theorem* (4.1.16) $f_i(X_i)$ is for each $i = 1, 2, 3, \ldots n$ a random variable. Let $B_1, B_2, \ldots B_n$ be Borel subsets of $\mathcal{R}$. We must show that the events $\{[f_i(X_i)]^{-1}(B_i) : i = 1, 2, \ldots n\}$ are mutually independent. Note that for each i

$$\begin{aligned}[f_i(X_i)]^{-1}(B_i) &= \{\omega : f_i(X_i(\omega)) \varepsilon B_i\} \\ &= \{\omega : X_i(\omega) \varepsilon f_i^{-1}(B_i)\} \\ &= X_i^{-1}(f_i^{-1}(B_i)).\end{aligned}$$

Since f_i is Borel measurable and B_i is a Borel set $f_i^{-1}(B_i)$ is a Borel set for each $i = 1, 2, 3, \ldots n$. Since $X_1, X_2, \ldots X_n$ are independent

$$\{X_i^{-1}(f_i^{-1}(B_i)) : i = 1, 2, \ldots n\} = \{[f_i(X_i)]^{-1}(B_i) : i = 1, 2, \ldots n\}$$

are mutually independent.

We may also obtain the following important generalization of the preceeding theorem:

*Theorem** 6.2.4

Let $1 \leq n_1 < n_2 < n_3 \cdots < n_k = n$. Let

$$f_1 : R^{n_1} \to R$$

and

$$f_i : R^{n_i - n_{i-1}} \to R \quad i = 2, 3, \ldots, k$$

be Borel measurable. (See *Note* (2.1.50)). If $X_1, X_2, \ldots, X_n$ are independent random variables, then the k random variables

$$f_1(X_1, X_2, \ldots, X_{n_1}), \; f_2(X_{n_1+1}, X_{n_1+2}, \ldots, X_{n_2}) \ldots$$

$$f_k(X_{n_{k-1}+1}, X_{n_{k-1}+2}, \ldots, X_{n_k})$$

are independent.

Theorem 6.2.5

Let X and Y be independent random variables. If $E[X]$ and $E[Y]$ exist then so does $E[XY]$ and furthermore

$$E[X \cdot Y] = E[X]E[Y].$$

If $E[XY]$ exists and neither X nor Y is zero P.a.e. then $E[X]$ and $E[Y]$ exist also and

$$E[XY] = E[X]E[Y].$$

Proof: The general method of proof shall be to prove that the theorem is true first in the case of simple functions X and Y. This fact together with the *Monotone Convergence Theorem* can then be used to extend the result to the case of non-negative random variables. By making use of the idea of the positive and negative parts of a random variable, the theorem is then extended to include general random variables X and Y.

Suppose that X and Y are simple functions with standard representations

$$X = \sum_{i=1}^{n} c_i \chi_{A_i} \quad \text{and} \quad Y = \sum_{j=1}^{m} d_j \chi_{B_j} .$$

Then

$$X \cdot Y = \left(\sum_{i=1}^{n} c_i \chi_{A_i}\right) \left(\sum_{j=1}^{m} d_j \chi_{B_j}\right)$$

$$= \sum_{i=1}^{n} \sum_{j=1}^{m} c_i d_j \chi_{A_i} \chi_{B_j}$$

$$= \sum_{i=1}^{n} \sum_{j=1}^{m} c_i d_j \chi_{A_i \cap B_j} .$$

Thus,

$$E[XY] = \int XY dP = \sum_{i=1}^{n} \sum_{j=1}^{m} c_i d_j P[A_i \cap B_j]$$

by *Corollary* (2.2.13). Since X and Y are independent, we have $A_i = X^{-1}\{c_i\}$ and $B_j = Y^{-1}\{d_j\}$ are independent and hence

$$P[A_i \cap B_j] = P[A_i]P[B_j]$$

by *Definition* (6.2.1). Thus,

$$E[XY] = \sum_{i=1}^{n} \sum_{j=1}^{m} c_i d_j \; P[A_i]P[B_j]$$

$$= \left(\sum_{i=1}^{n} c_i P[A_i] \right) \left(\sum_{j=1}^{m} d_j \; P[B_j] \right)$$

$$= \int X dP \int Y dP$$

$$= E[X]E[Y].$$

Note that the expectation always exists for a simple function. Now let X and Y be arbitrary non-negative independent random variables. Define

$$X_n(\omega) = \begin{cases} \dfrac{i}{2^n} & \text{if } \dfrac{i}{2^n} \le X(\omega) < \dfrac{i+1}{2^n}, \; i = 1, 2, \ldots, n2^n \\ 0, & \text{otherwise} \end{cases}$$

and define Y_n similarly in terms of Y. X_n and Y_n are obviously simple random variables. They are also independent as we shall now show.

Let

$$\phi_n = X\chi_{X^{-1}\left[0, \frac{n2^n + 1}{2^n}\right]}$$

and let

$$\psi_n = Y\chi_{Y^{-1}[0, \frac{n2^n + 1}{2^n}]}.$$

Let B_1 and B_2 be Borel sets. Note that

$$\begin{aligned}
\phi_n^{-1}(B_1) &= \{\omega : \phi_n(\omega) \in B_1\} \\
&= \{\omega : X\chi_{X^{-1}[0, \frac{n2^n + 1}{2^n}]}(\omega) \in B_1\} \\
&= \{\omega : X(\omega) \in B_1 \text{ and } \omega \in X^{-1}[0, \frac{n2^n + 1}{2^n}]\}. \\
&= X^{-1}(B_1) \cap X^{-1}[0, \frac{n2^n + 1}{2^n}] \\
&= X^{-1}(B_1 \cap [0, \frac{n2^n + 1}{2^n}]), \quad \text{by } \textit{Theorem } (1.2.13).
\end{aligned}$$

Similarly,

$$\psi_n^{-1}(B_2) = Y^{-1}(B_2 \cap [0, \frac{n2^n + 1}{2^n}]).$$

Note that

$$B_1 \cap [0, \frac{n2^n + 1}{2^n}] \quad \text{and} \quad B_2 \cap [0, \frac{n2^n + 1}{2^n}]$$

are Borel sets and hence since X and Y are independent the events

$$\phi_n^{-1}(B_1) = X^{-1}(B_1 \cap [0, \frac{n2^n + 1}{2^n}])$$

and

$$\psi_n^{-1}(B_2) = Y^{-1}(B_2 \cap [0, \frac{n2^n + 1}{2^n}])$$

are independent by *Definition* (6.2.1). Thus, ϕ_n and ψ_n are independent random variables.

Note that X_n and Y_n may be expressed in the form

$$X_n = \frac{[2^n \cdot \phi_n]}{2^n}$$

$$Y_n = \frac{[2^n \cdot \psi_n]}{2^n}$$

where $[\cdot]$ denotes the greatest integer function. Note that the greatest integer function is an increasing function and hence by *Theorem* (2.1.43) it is Borel measurable. Hence, by *Theorem* (6.2.3), X_n and Y_n are independent. Observe also that if

$$X_n(\omega) = \frac{i}{2^n},$$

then

$$X(\omega) \geq \frac{i}{2^n} = \frac{2i}{2^{n+1}} \quad \text{where} \quad 0 \leq i < n2^n.$$

Since

$$0 \leq 2i < n2^{n+1} < (n + 1)2^{n+1},$$

$$X(\omega) \geq \frac{2i}{2^{n+1}}$$

implies that

$$X_{n+1}(\omega) \geq \frac{2i}{2^{n+1}} = \frac{i}{2^n} = X_n(\omega).$$

Thus, we have $0 \leq X_n \leq X_{n+1} \leq X$. Similarly, $0 \leq Y_n \leq Y_{n+1} \leq Y$. It can be shown as follows that

$$\lim_{n\to\infty} X_n = X \quad \text{and} \quad \lim_{n\to\infty} Y_n = Y.$$

Choose $\varepsilon > 0$. There exists a positive integer N_1 such that $n \geq N_1$ implies that $\frac{1}{2^n} < \varepsilon$. Let $\omega \in \Omega$ be arbitrary but fixed. Since X is a non-negative function, there exists a positive integer N_ω such that

$$X(\omega) < N_\omega + \frac{1}{2^{N_\omega}}.$$

Let $N_2 = \max(N_1, N_\omega)$. Note that since (X_n) is an increasing sequence we have from the definition of X_n that for $n \geq N_2$

$$|X(\omega) - X_n(\omega)| < \frac{1}{2^{N_2}} < \varepsilon .$$

Hence,

$$\lim_{n\to\infty} X_n = X.$$

A similar argument will show that $\lim_{n\to\infty} Y_n = Y$. We thus have the hypothesis of the *Monotone Convergence Theorem* (2.2.26) satisfied and may conclude that

$$\lim_{n\to\infty} \int X_n \, dP = \int X dP \qquad (1)$$

$$\lim_{n\to\infty} \int Y_n \, dP = \int Y dP \qquad (2)$$

and

$$\lim_{n\to\infty} \int X_n Y_n \, dP = \int XY dP \qquad (3).$$

If (1) and (2) are finite, we have, since X_n and Y_n are simple independent random variables that

$$E[X]E[Y] = \lim_{n\to\infty} E[X_n]E[Y_n] = \lim_{n\to\infty} E[X_n Y_n] \quad \text{is finite.}$$

We have from equation (3) that

$$\lim_{n\to\infty} E[X_n Y_n] = \int XY dP \quad \text{is finite.}$$

Hence

$$E[XY] = \int XY dP$$

exists and furthermore

$$E[XY] = E[X]E[Y].$$

Conversely, if (3) is finite, then we have

$$\lim_{n\to\infty} E[X_n Y_n] = \lim_{n\to\infty} E[X_n]E[Y_n] = E[XY] \quad \text{is finite.}$$

Hence $\lim_{n\to\infty} E[X_n] = E[X]$ and $\lim_{n\to\infty} E[Y_n] = E[Y]$ are finite since neither X nor Y is zero P.a.e. (See *Note* (2.1.4)). Thus again we have

$$E[XY] = E[X]E[Y]$$

and the theorem is proved for non-negative random variables. Now let X and Y represent arbitrary independent random variables. Note that

$$X = X_+ - X_-$$

and

$$Y = Y_+ - Y_-. \quad \text{(See } \textit{Theorem} \text{ (2.2.31))}.$$

It can be shown that

X_+ and Y_+ are independent;

X_+ and Y_- are independent;

X_- and Y_+ are independent;

X_- and Y_- are independent.

Thus if $E[XY]$ exists,

$$\begin{aligned} E[XY] &= E[(X_+ - X_-)(Y_+ - Y_-)] \\ &= E[X_+Y_+ - X_+Y_- - X_-Y_+ + X_-Y_-] \\ &= E[X_+Y_+] - E[X_+Y_-] - E[X_-Y_+] + E[X_-Y_-] \\ &= E[X_+]E[Y_+] - E[X_+]E[Y_-] - E[X_-]E[Y_+] + E[X_-]E[Y_-] \\ &= [E[X_+] - E[X_-]][E[Y_+] - E[Y_-]] \\ &= E[X]E[Y]. \end{aligned}$$

Similarly, if $E[X]$ and $E[Y]$ exist, then $E[XY]$ exists and

$$E[XY] = E[X]E[Y].$$

The following theorem provides an alternative approach to the idea of stochastic independence. This theorem is in fact often taken as the definition of stochastic independence.

*Theorem** 6.2.6

Let $\{X_\gamma : \gamma \in \Gamma\}$ be a collection of one dimensional random variables defined relative to the probability space $(\Omega, \mathcal{J}, P)$. These random variables are stochastically independent if and only if for every natural number n and every n distinct elements $\gamma_1, \gamma_2, \ldots, \gamma_n \in \Gamma$ we have

$$H_{X_{\gamma_1} X_{\gamma_2} \cdots X_{\gamma_n}}(x_1, x_2, \ldots, x_n) = \prod_{j=1}^{n} F_{X_{\gamma_j}}(x_j)$$

for all $(x_1, x_2, \ldots x_n) \in R^n$ where $H_{X_{\gamma_1} X_{\gamma_2} \cdots X_{\gamma_n}}$ represents the joint distribution function of the n-dimensional random variable $(X_{\gamma_1}, X_{\gamma_2}, \ldots, X_{\gamma_n})$ and $F_{X_{\gamma_j}}$ represents the distribution function for the one dimensional random variable X_{γ_j}.

Definition 6.2.7

Let $\{X_\gamma : \gamma \in \Gamma\}$ be a collection of one dimensional random variables defined relative to the probability space $(\Omega, \mathcal{J}, P)$. These random variables are said to be *pairwise independent* if for each pair $\gamma_1, \gamma_2, \in \Gamma$ such that $\gamma_1 \neq \gamma_2$ we have

$$H_{X_{\gamma_1} X_{\gamma_2}}(x_1, x_2) = F_{X_{\gamma_1}}(x_1) F_{X_{\gamma_2}}(x_2)$$

where $H_{X_{\gamma_1} X_{\gamma_2}}$, $F_{X_{\gamma_1}}$ and $F_{X_{\gamma_2}}$ are as in *Theorem* (6.2.6).

Theorem 6.2.8

Let $\{X_\gamma : \gamma \in \Gamma\}$ be a collection of stochastically independent one-dimensional random variables. $\{X_\gamma : \gamma \in \Gamma\}$ is also a collection of pairwise independent random variables.

The following example will show that the converse of *Theorem* (6.2.8) is false. That is, pairwise independence does not imply stochastic independence.

Example 6.2.9

Consider the experiment of tossing a fair coin twice and observing the outcome. Denote a possible outcome of this experiment by an ordered pair of the form (H, T) where H represents heads and T tails. Let

$\Omega = \{\omega_1, \omega_2, \omega_3, \omega_4\}$ where

$\omega_1 = (H, H)$ $\qquad$ $\omega_3 = (T, H)$

$\omega_2 = (H, T)$ $\qquad$ $\omega_4 = (T, T)$.

Define a set function P on $\mathcal{S}$ where $\mathcal{S} = \rho_\Omega$ by $P\{\omega_i\} = 1/4$ $\quad i = 1, 2, 3, 4$; for $E \in \mathcal{S}$, $E \neq \phi$ $P[E] = \sum_{\omega \in E} P\{\omega\}$; $P[\phi] = 0$. $(\Omega, \mathcal{S}, P)$ is a probability space. Consider the events

$$E_1 = \{\omega_1, \omega_2\}$$
$$E_2 = \{\omega_1, \omega_3\}$$
$$E_3 = \{\omega_2, \omega_3\}.$$

The indicator functions χ_{E_1}, χ_{E_2} and χ_{E_3} from Ω into $\{0, 1\}$ are one-dimensional random variables. $\chi_{E_1}, \chi_{E_2}, \chi_{E_3}$ is not a collection of stochastically independent random variables but it is a collection of pairwise independent random variables.

Verification: It is obvious that $(\Omega, \mathcal{J}, P)$ is a probability space and that χ_{E_1}, χ_{E_2} and χ_{E_3} are one-dimensional random variables. To see that they are not stochastically independent, note that

$$\begin{aligned} H_{\chi_{E_1}\chi_{E_2}\chi_{E_3}}(0, 0, 0) &= P[\chi_{E_1} \leq 0, \chi_{E_2} \leq 0, \chi_{E_3} \leq 0] \\ &= P\{\omega_4\} = 1/4. \end{aligned}$$

However,

$$F_{\chi_{E_1}}(0) = P[\chi_{E_1} \leq 0] = P\{\omega_3, \omega_4\} = 1/2$$
$$F_{\chi_{E_2}}(0) = P[\chi_{E_2} \leq 0] = P\{\omega_2, \omega_4\} = 1/2$$
$$F_{\chi_{E_3}}(0) = P[\chi_{E_3} \leq 0] = P\{\omega_1, \omega_4\} = 1/2.$$

Hence,

$$F_{\chi_{E_1}}(0)\, F_{\chi_{E_2}}(0)\, F_{\chi_{E_3}}(0) = 1/2 \cdot 1/2 \cdot 1/2 = 1/8 \neq 1/4 = H(0, 0, 0).$$

To show that χ_{E_1}, χ_{E_2} and χ_{E_3} are pairwise independent, note first that the events E_1' and E_2' are independent since $P[E_1' \cap E_2'] = P\{\omega_4\} = 1/4 = 1/2 \cdot 1/2 = P[E_1']P[E_2']$. Let $(x_1, x_2) \in R^2$ and consider

$$H_{\chi_{E_1} \chi_{E_2}}(x_1, x_2) = P[\chi_{E_1} \le x_1, \chi_{E_2} \le x_2]$$

$$= P[\{\omega : \chi_{E_1}(\omega) \le x_1\} \cap \{\omega : \chi_{E_2}(\omega) \le x_2\}].$$

Note that $\{\omega : \chi_{E_1}(\omega) \le x_1\}$ is either ϕ, E_1' or Ω depending on whether x_1 satisfies the condition $x_1 < 0$, $0 \le x_1 < 1$, or $1 \le x_1$, respectively. Similarly $\{\omega : \chi_{E_2}(\omega) \le x_2\}$ is either ϕ, E_2' or Ω depending on the value of x_2. Hence, it is necessary to consider the following possible situations:

$\{\omega : \chi_{E_1}(\omega) \le x_1\}$ \ $\{\omega : \chi_{E_2}(\omega) \le x_2\}$	ϕ	E_2'	Ω
ϕ	ϕ	ϕ	ϕ
E_1'	ϕ	$\{\omega_4\}$	E_1'
Ω	ϕ	E_2'	Ω

If $\{\omega : \chi_{E_1}(\omega) \leq x_1\} \cap \{\omega : \chi_{E_2}(\omega) \leq x_2\} = \phi$, then $H_{\chi_{E_1}\chi_{E_2}}(x_1, x_2) = P[\phi] = 0$. From the above table it can be seen that this implies that at least one event in question must be the null event. Without loss of generality let us assume that $\{\omega : \chi_{E_1}(\omega) \leq x_1\} = \phi$ implying that $F_{\chi_{E_1}}(x_1) = P\{\omega : \chi_{E_1}(\omega) \leq x_1\} = P[\phi] = 0$. Hence, $F_{\chi_{E_1}}(x_1) F_{\chi_{E_2}}(x_2) = 0 = H_{\chi_{E_1}\chi_{E_2}}(x_1, x_2)$ as desired.

Similar reasoning shows that if $\{\omega : \chi_{E_1}(\omega) \leq x_1\} \cap \{\omega : \chi_{E_2}(\omega) \leq x_2\} = \Omega$ then both of the above events are equal to Ω and in this case we have $H_{\chi_{E_1}\chi_{E_2}}(x_1, x_2) = 1 = 1 \cdot 1 = F_{\chi_{E_1}}(x_1) F_{\chi_{E_2}}(x_2)$.

If $\{\omega : \chi_{E_1}(\omega) \leq x_1\} \cap \{\omega : \chi_{E_2}(\omega) \leq x_2\} = E_2'$ then from the above table,

$$\{\omega : \chi_{E_1}(\omega) \leq x_1\} = \Omega \text{ and } \{\omega : \chi_{E_2}(\omega) \leq x_2\} = E_2'.$$

Thus,

$$H_{\chi_{E_1}\chi_{E_2}}(x_1, x_2) = P[E_2'] = 1 \cdot P[E_2'] = F_{\chi_{E_1}}(x_1) F_{\chi_{E_2}}(x_2).$$

A similar argument holds if

$$\{\omega : \chi_{E_1}(\omega) \leq x_1\} \cap \{\omega : \chi_{E_2}(\omega) \leq x_2\} = E_1'.$$

The only possibility left to consider is that in which

$$\{\omega : \chi_{E_1}(\omega) \le x_1\} \cap \{\omega : \chi_{E_2}(\omega) \le x_2\} = \{\omega_4\}.$$

In this case

$$\begin{aligned} H_{\chi_{E_1}\chi_{E_2}}(x_1, x_2) &= P[E_1' \cap E_2'] \\ &= P[E_1']P[E_2'] \text{ , due to independence of the events } E_1' \text{ and } E_2' \\ &= P\{\omega : \chi_{E_1}(\omega) \le x_1\}P\{\omega : \chi_{E_2}(\omega) \le x_2\} \\ &= F_{\chi_{E_1}}(x_1)\, F_{\chi_{E_2}}(x_2). \end{aligned}$$

The proof that the pairs χ_{E_1}, χ_{E_3} and χ_{E_2}, χ_{E_3} satisfy analogous conditions is identical to that given above. Hence $\{\chi_{E_1}, \chi_{E_2}, \chi_{E_3}\}$ satisfy the requirements of *Definition* (6.2.7).

6.2 *EXERCISES*

1. Let $\{X_\gamma : \gamma \in \Gamma\}$ be a collection of pairwise independent random variables. Prove that for any $n \ge 2$ and any collection $\{\gamma_i : i = 1, 2, \ldots n\} \subseteq \Gamma$, $\text{Var} \sum_{i=1}^{n} X_{\gamma_i} = \sum_{i=1}^{n} \text{Var } X_{\gamma_i}$.

The following collection of exercises deals with a class of theorems called the "weak laws of large numbers." These theorems generally concern sequences of random variables (X_n) and (Y_n) where $Y_n = \frac{\sum_{i=1}^{n} X_i}{n}$.

They deal with conditions under which (Y_n) *converges in probability* to a specified constant. The reason for the term "weak" law as opposed to a so called "strong" law is quite simple. The classical "strong" laws of large numbers deal with convergence with probability one. As seen in Chapter Five convergence with probability one is a stronger mode of convergence than is convergence in probability. For this reason, theorems involving the former mode are referred to as "strong laws" and those involving the latter as "weak laws."

2. *(Tchebychef's Theorem)*

   Let (X_n) be a sequence of pairwise independent random variables such that $\mathrm{Var}(X_i) < c$ for each $i = 1, 2, \ldots$. Then for $\alpha > 0$

$$\lim_{n\to\infty} P\{\omega : |\frac{\sum_{i=1}^{n} X_i}{n} - \frac{1}{n}\sum_{i=1}^{n} E[X_i]| < \alpha\} = 1.$$

There are several weak laws which are direct consequences of Tchebychef's Theorem. Notable among them is the Bernoulli Theorem which was one of the first limit theorems ever proved in the development of probability theory. It was discovered by Jakob Bernoulli and published in 1715 after his death. Its proof is based on the following exercise.

3. Let (X_n) be a sequence of pairwise independent point binomial random variables with parameters p_n. Then for $\alpha > 0$

$$\lim_{n\to\infty} P\{\omega : |\frac{\sum_{i=1}^{n} X_i}{n} - \frac{1}{n}\sum_{i=1}^{n} p_i| < \alpha\} = 1.$$

Consider an experiment which is to be performed once; which can result in one of two possible outcomes one to be considered as "success" and the other as "failure"; the probability of success being p. In this situation we can define a random variable X by

$$X = \begin{cases} 1 & \text{if a success is obtained} \\ 0 & \text{if a failure occurs.} \end{cases}$$

The density f for X is given by

$$f(x) = \begin{cases} p^x (1-p)^{1-x} & , \quad x = 0, 1 \\ 0 & , \quad \text{elsewhere.} \end{cases}$$

By *Example* (4.2.10) X is nothing more than a point binomial random variable with parameter p. Note that the random variable $Z_n = X_1 + X_2 + \ldots X_n$ where for each i, X_i is point binomial with parameter p_i can be thought of physically as giving the number of success in a set of n independent trials where the probability of success on the ith trial is p_i. If for each i, $p_i = p$, then Z_n is a binomial random variable with parameters n and p. These facts allow us to obtain Poisson's theorem directly from *Exercise* (6.3.3) and to obtain Bernoulli's Theorem as a special case of Poisson's Theorem.

4. *(Poisson's Theorem)*

Let Z_n be the number of successes obtained in a set of n independent trials where the probability of success on the ith trial is p_i. Then for $\alpha > 0$,

$$\lim_{n\to\infty} P\{\omega : \left| \frac{Z_n}{n} - \frac{1}{n}\sum_{i=1}^{n} p_i \right| < \alpha\} = 1.$$

5. *(Bernoulli's Theorem)*

Let Z_n be the number of successes obtained in a set of n independent trials where the probability of success on each trial is p. Then for $\alpha > 0$,

$$\lim_{n\to\infty} P\{\omega : \left| \frac{Z_n}{n} - p \right| < \alpha\} = 1.$$

The following exercise serves as a basis for the "rule of the arithmetic mean" which is in constant use in the theory of measurement. Suppose that some physical quantity a is being measured. Repeating the measurement n times under identical conditions results in n observations $x_1, x_2, \ldots, x_n$ which do not completely coincide and which can be considered to be n realizations on the random variables $X_1, X_2, \ldots, X_n$. It is natural to use as an estimator for a the random variable $\dfrac{\sum_{i=1}^{n} X_i}{n}$. Each of the random variables $X_1, X_2, \ldots, X_n$ should have expectation a. According to the following exercise we can obtain a value as close to the required quantity a as we wish by making the number of measurements sufficiently large.

6. Let (X_n) be a sequence of pairwise independent random variables such that $E[X_i] = a$ and $\mathrm{Var}(X_i) < c$ for each i. Then for $\alpha > 0$

$$\lim_{n\to\infty} P\{\omega : \left|\frac{\sum_{i=1}^{n} X_i}{n} - a\right| < \alpha\} = 1.$$

6.3 *SUMMARY*

The concept of an *n-dimensional random variable* was introduced with the following definition:

Let $(\Omega, \mathcal{S}, P)$ be a probability space. Let $X : \Omega \to \mathcal{R}^n$. X is said to be an *n-dimensional random variable* if and only if X is measurable.

n-dimensional discrete and *n-dimensional continuous random variables* were distinguished in a manner analogous to the one dimensional case.

Independence of random variables was defined in a manner which paralleled the idea of independent events presented in Chapter Three.

Let $K = \{X_\gamma : \gamma \in \Gamma\}$ be a collection of one-dimensional random variables relative to the probability space $(\Omega, \mathcal{S}, P)$. These random variables are said to be *stochastically independent* if for every natural number n and every n distinct elements $\gamma_1, \gamma_2, \ldots, \gamma_n \in \Gamma$ and any collection $\{B_1, B_2, \ldots, B_n\}$ of Borel subsets of $\mathcal{R}$ the events

$$\{X_{\gamma_1}^{-1}(B_1), X_{\gamma_2}^{-1}(B_2), \ldots, X_{\gamma_n}^{-1}(B_n)\}$$

are *mutually independent.*

A series of theorems were presented which summarized some of the more important implications of the independence property. The chapter was concluded by a discussion of the idea of *pairwise independence* and the relationship between *pairwise independence* and *stochastic independence* was considered.

IMPORTANT TERMS IN CHAPTER SIX

n-*dimensional random variable*

joint distribution function

n-*dimensional discrete random variable*

joint density function (discrete and continuous)

n-*dimensional continuous random variable*

stochastic independence

pairwise independence

REFERENCES AND SUGGESTIONS FOR FURTHER READINGS

[1] **Burrill, C. W., Measure, Integration and Probability. New York: McGraw-Hill Book Company, 1972.

[2] Gnedenko, B. V., The Theory of Probability. New York: Chelsea Publishing Company, 1966.

[3] **Tucker, H. G., A Graduate Course in Probability. New York: Academic Press, 1967.

**These books are more advanced than the approach of the present text.

CHAPTER SEVEN

SOME LIMIT THEOREMS

7.0 *INTRODUCTION*

In this chapter we shall investigate several of the most widely used of the so called limit theorems. We shall begin with the classical inequality derived by Tchebychef (See *Exercise* (6.2.2)) which justifies the role that the standard deviation plays as a measure of variability. Following this discussion, we shall present *Markov's* inequality and the *Weak Law of Large Numbers*. We shall continue our discussion into a more general form of the latter law, namely, *Kolmogorov's Strong Law of Large Numbers*.

The remainder of the chapter is devoted to the *Central Limit Theorem*, which is one of the most important theorems in the whole of mathematics. That is, in a sequence of stochastically independent random variables, $X_1, X_2, \ldots, X_n$, their sum, for sufficiently large n, is *approximately normally distributed* under certain conditions. We shall confine our study to two versions of the Central Limit Theorem, namely, when the random variables in the sequence are independent and identically distributed and when the random variables are independent but not identically distributed.

The Central Limit Theorem was formulated by Laplace and Gauss in the early 19th century. However, no formal presentation was given until 1901, when Laplace gave a rigorous mathematical proof of the theorem.

7.1 *LIMIT THEOREMS*

We shall begin our study of limit theorems with the *General Tchebychef's Inequality.*

Theorem 7.1.1

(*General Tchebychef's Inequality*)

Let X be a random variable and let $f : R \to R$ be Borel measurable. Assume that $E[f(X)]$ exists. Then

i) If f is an even non-negative function which is non-decreasing on $[0, \infty)$ then for every $\lambda > 0$ the inequality

$$P\{\omega : |X(\omega)| \geq \lambda\} \leq \frac{1}{f(\lambda)} E[f(X)]$$

holds whenever $f(\lambda) > 0$;

ii) If f is a non-decreasing, non-negative function on $(-\infty, \infty)$ then

$$P\{\omega : X(\omega) \geq \lambda\} \leq \frac{1}{f(\lambda)} E[f(X)]$$

holds for all real numbers λ provided $f(\lambda) > 0$.

Proof: We shall prove case i). The proof of case ii) is analogous and is therefore left to the reader.

i) Note that since f is even and non-decreasing on $[0, \infty)$ it follows that whenever

$$|x| \leq \lambda, \quad f(x) \leq f(\lambda).$$

Thus by *Theorem* (4.3.16),

$$E[f(X)] = \int f(X)dP = \int_{-\infty}^{\infty} f(x)dF_X(x).$$

However,

$$\begin{aligned}
\int_{-\infty}^{\infty} f(x)dF_X(x) &\geq \int_{-\infty}^{-\lambda} f(x)dF_X(x) + \int_{\lambda}^{\infty} f(x)dF_X(x) \\
&\geq \int_{-\infty}^{-\lambda} f(-\lambda)dF_X(x) + \int_{\lambda}^{\infty} f(\lambda)dF_X(x) \\
&= f(-\lambda)\int_{-\infty}^{-\lambda} dF_X(x) + f(\lambda)\int_{\lambda}^{\infty} dF_X(x) \\
&= f(-\lambda)P\{\omega : X(\omega) \leq -\lambda\} + f(\lambda)P\{\omega : X(\omega) \geq \lambda\}, \\
&\qquad\qquad \text{(See } \textit{Note} \text{ (4.2.24))} \\
&= f(\lambda)[P\{\omega : X(\omega) \leq -\lambda\} + P\{\omega : X(\omega) \geq \lambda\}] \\
&= f(\lambda)P\{\omega : |X(\omega)| \geq \lambda\}.
\end{aligned}$$

Thus,

$$P\{\omega : |X(\omega)| \geq \lambda\} \leq \frac{1}{f(\lambda)} E[f(X)]$$

as was desired.

Corollary 7.1.2

(*Markov Inequality*)

Let X be a random variable and let $\alpha \geq 0$. Assume that $E[|X|^\alpha]$ exists. Then for $\lambda > 0$

$$P\{\omega : |X(\omega)| \geq \lambda\} \leq \frac{1}{\lambda^\alpha} E[|X|^\alpha].$$

Corollary 7.1.3

(*Tchebychef's Inequality*)

Let X be a random variable such that $E[X] = \mu$ and $Var(X) = \sigma^2$ exist. Then for any real number $k > 0$

$$P\{\omega : |X(\omega) - \mu| \geq k\sigma\} \leq \frac{1}{k^2}.$$

Theorem 7.1.4

(*The Weak Law of Large Numbers*)

Let $X_1, X_2, X_3, \ldots$ be a sequence of identically distributed and independent random variables such that $E[X_i] = \mu$ and $Var[X_i] = \sigma^2$ exist. Define a sequence $S_1, S_2, S_3, \ldots$ of random variables by

$$S_n = \sum_{i=1}^{n} X_i.$$

Then for each $\varepsilon > 0$,

$$\lim_{n\to\infty} P\{\omega : |\frac{S_n(\omega)}{n} - \mu| \geq \varepsilon\} = 0.$$

Proof: Consider the random variable $\frac{S_n}{n} - \mu$ where n is arbitrary but fixed.

$$E\left[\frac{S_n}{n} - \mu\right] = \frac{1}{n} E[S_n] - \mu$$

$$= \frac{1}{n} E\left[\sum_{i=1}^{n} X_i\right] - \mu$$

$$= \frac{1}{n} \cdot (n\,\mu) - \mu = 0.$$

$$\mathrm{Var}\left[\frac{S_n}{n} - \mu\right] = E\left[\left(\frac{S_n}{n} - \mu\right)^2\right]$$

$$= E\left[\frac{1}{n^2} S_n^{\,2} - \frac{2\mu}{n} S_n + \mu^2\right]$$

$$= \frac{1}{n^2} E[S_n^{\,2}] - \frac{2\mu}{n} E[S_n] + \mu^2$$

$$= \frac{1}{n^2} E[S_n^{\,2}] - \mu^2$$

$$= \frac{1}{n^2} \left[E\left[\sum_{i=1}^{n} X_i\right]^2\right] - \mu^2$$

$$= \frac{1}{n^2} E\left[\sum_{i=1}^{n} X_i^{\,2} + 2 \sum_{\substack{i,j \\ i\neq j}} X_i X_j\right] - \mu^2$$

$$= \frac{1}{n^2} \left[n\, E[X_i^{\,2}] + 2 \sum_{\substack{i,j \\ i\neq j}} E[X_i X_j]\right] - \mu^2$$

Note that for $i \neq j$, X_i and X_j are independent. Thus, by *Theorem* (6.2.5), $E[X_i X_j] = E[X_i]E[X_j] = \mu^2$. Hence,

$$\mathrm{Var}\left[\frac{S_n}{n} - \mu\right] = \frac{1}{n^2}\left[n\, E[X_i^{\,2}] + 2 \sum_{\substack{i,j \\ i\neq j}} \mu^2\right] - \mu^2$$

$$= \frac{1}{n} E[X_i^{\,2}] + \frac{2}{n^2} \binom{n}{2} \mu^2 - \mu^2$$

$$= \frac{1}{n} E[X_i^2] + \frac{2}{n^2} \frac{n(n-1)}{2} \mu^2 - \mu^2$$

$$= \frac{1}{n} E[X_i^2] + (1 - \frac{1}{n})\mu^2 - \mu^2$$

$$= \frac{1}{n} E[X_i^2] - \frac{1}{n} \mu^2$$

$$= \frac{1}{n} [E[X_i^2] - \mu^2]$$

$$= \frac{1}{n} \text{Var } X_i$$

$$= \frac{\sigma^2}{n}$$

Thus,

$$E[\frac{S_n}{n} - \mu] = 0 \quad \text{and} \quad \text{Var}[\frac{S_n}{n} - \mu] = \frac{\sigma^2}{n}$$

Applying *Corollary* (7.1.3) to $\frac{S_n}{n} - \mu$, we have that

$$P\{\omega : |\frac{S_{n(\omega)}}{n} - \mu| \geq \varepsilon\} \leq \frac{1}{\left(\frac{\sqrt{n}\,\varepsilon}{\sigma}\right)^2} = \frac{\sigma^2}{n\varepsilon^2} .$$

Thus,

$$\lim_{n\to\infty} P\{\omega : |\frac{S_{n(\omega)}}{n} - \mu| \geq \varepsilon\} \leq \lim_{n\to\infty} \frac{\sigma^2}{\varepsilon^2} \frac{1}{n} = 0.$$

Note 7.1.5

The weak law of large numbers derives its name from the fact that it in essence states that the sequence $\left(\frac{S_n}{n}\right)$ converges to μ in probability. This type of convergence is weak convergence in the sense that it does not imply convergence almost uniformly or convergence in mean square or convergence with probability one but it is itself implied by each of the above.

Theorem 7.1.6

(Kolmogorov's Inequality)

Let $X_1, X_2, X_3, \ldots, X_n$ be independent random variables such that

$$\mathrm{Var}[X_i] = \sigma_i^2 \quad \text{exists for} \quad i = 1, 2, \ldots n.$$

Let $S_j = \sum_{i=1}^{j} [X_i - E[X_i]]$, $j = 1, 2, \ldots n$. Then for every $\varepsilon > 0$ we have

$$P\{\omega : \max_{1 \le j \le n} |S_j(\omega)| \ge \varepsilon\} \le \frac{\mathrm{Var}[S_n]}{\varepsilon^2}.$$

Proof: Let $\varepsilon > 0$ be given and consider the following events:

$$A = \{\omega : \max_{1 \le j \le n} |S_j(\omega)| \ge \varepsilon\}$$

$$A_j = \{\omega : \max_{1 \le i < j} |S_i(\omega)| < \varepsilon, |S_j(\omega)| \ge \varepsilon\}.$$

Note that

$$A_i \cap A_j = \phi \quad i \ne j$$

and that

$$\bigcup_{j=1}^{n} A_j = A.$$

Now

$$E[S_j] = E[\sum_{i=1}^{j} [X_i - E[X_i]]]$$

$$= \sum_{i=1}^{j} E[X_i - E[X_i]] = 0.$$

In particular,

$$E[S_n] = 0$$

implying that

$$\mathrm{Var}[S_n] = E[(S_n - E[S_n])^2] = E[S_n^2] .$$

Note that $S_n^2 \geq S_n^2 \chi_A$ and hence

$$E[S_n^2] \geq E[S_n^2 \chi_A] = E[S_n^2 \sum_{i=1}^{n} \chi_{A_i}]$$

$$= E[\sum_{i=1}^{n} S_n^2 \chi_{A_i}]$$

$$= \sum_{i=1}^{n} E[S_n^2 \chi_{A_i}]$$

where χ_A and χ_{A_i} are the usual indicator functions.

Thus,

(1) $$\mathrm{Var}[S_n] \geq \sum_{i=1}^{n} E[S_n^2 \chi_{A_i}].$$

Now

$$E[S_n^2 \chi_{A_i}] = E[(S_i + (S_n - S_i))^2 \chi_{A_i}]$$

$$= E[S_i^2 \chi_{A_i}] + 2E[S_i(S_n - S_i)\chi_{A_i}]$$

$$+ E[(S_n - S_i)^2 \chi_{A_i}].$$

Consider the term

$$2E[S_i(S_n - S_i)\chi_{A_i}].$$

Note that the random variable $S_i\chi_{A_i}$ is a function of $X_1, X_2, \ldots X_i$ and $S_n - S_i$ is a function of $X_{i+1}, X_{i+2}, \ldots, X_n$. By *Theorem* (6.2.4) $S_i\chi_{A_i}$ and $S_n - S_i$ are independent. Since $E[S_i\chi_{A_i}]$ and $E[S_n - S_i]$ exist, we have by *Theorem* (6.2.5) that

$$E[S_i\chi_{A_i}(S_n - S_i)]$$

exists and

$$E[S_i\chi_{A_i}(S_n - S_i)] = E[S_i\chi_{A_i}]E[S_n - S_i].$$

However,

$$E[S_n - S_i] = E[S_n] - E[S_i] = 0$$

and thus we have

$$E[S_n^2 \chi_{A_i}] = E[S_i^2 \chi_{A_i}] + E[(S_n - S_i)^2 \chi_{A_i}].$$

(2) $$E[S_i^2 \chi_{A_i}] + E[(S_n - S_i)^2 \chi_{A_i}] \geq E[S_i^2 \chi_{A_i}]$$
$$\geq E[\varepsilon^2 \chi_{A_i}]$$
$$= \int \varepsilon^2 \chi_{A_i}\, dP$$
$$= \varepsilon^2 P[A_i].$$

Combining (1) and (2) we have

$$\text{Var}[S_n] \geq \sum_{i=1}^{n} \varepsilon^2 P[A_i]$$
$$= \varepsilon^2 P[A].$$

Thus

$$P[A] \leq \frac{\text{Var}[{}^S n]}{\varepsilon^2}$$

as was to be proved.

Theorem 7.1.7

Let $X_1, X_2, \ldots$ be a sequence of independent random variables with finite variances. If

$$\sum_{i=1}^{\infty} \text{Var}[X_i] < \infty$$

then

$$\sum_{i=1}^{\infty} [X_i - E[X_i]]$$

converges P.a.e.

Proof: We shall apply the Cauchy Criterion for convergence, *Theorem* (2.4.13). Define S_n, T_n and T by

$$S_n = \sum_{i=1}^{n} [X_i - E[X_i]];$$

$$T_n = \sup\{|S_{n+k} - S_n| : k = 1, 2, 3, \ldots\};$$

$$T = \inf\{T_n : n = 1, 2, 3, \ldots\}.$$

If $T(\omega) = 0$, then for each $\varepsilon > 0$ there exists an integer $N_1 \geq 1$ such that

$$T_{N_1}(\omega) < \varepsilon.$$

Thus, from the definition of T_{N_1} we have

$$|S_{N_1+k}(\omega) - S_{N_1}(\omega)| < \varepsilon \quad \text{for} \quad k = 1, 2, 3, \ldots .$$

Thus, if $T(\omega) = 0$, the sequence $(S_n(\omega))$ is a Cauchy sequence and therefore converges. We shall show that $T = 0$ P.a.e. which will in turn imply that (S_n) converges P.a.e. Since S_n is nothing more than the nth partial sum of the series of interest, the proof will be complete. By the Kolmogorov Inequality, *Theorem* (7.1.6), we have that for all $n \geq 1$ and any natural number N_1

$$P\{\omega : \max_{1 \leq k \leq n} |S_{N_1+k}(\omega) - S_{N_1}(\omega)| \geq \varepsilon\} \leq \frac{1}{\varepsilon^2} \operatorname{Var}[S_{N_1+n} - S_{N_1}]$$

$$\leq \frac{1}{\varepsilon^2} \operatorname{Var}[\sum_{j=N_1+1}^{N_1+n} (X_j - E[X_j])]$$

$$= \frac{1}{\varepsilon^2} \operatorname{Var}[\sum_{j=N_1+1}^{N_1+n} X_j]$$

$$= \frac{1}{\varepsilon^2} \sum_{j=N_1+1}^{N_1+n} \operatorname{Var}[X_j]$$

$$\leq \frac{1}{\varepsilon^2} \sum_{j=N_1+1}^{\infty} \operatorname{Var}[X_j] .$$

Hence,

$$P\{\omega : T_{N_1}(\omega) \geq \varepsilon\} = P\{\omega : \sup_{k \geq 1} |S_{N_1+k}(\omega) - S_{N_1}(\omega)| \geq \varepsilon\}$$

$$\leq \frac{1}{\varepsilon^2} \sum_{j=N_1+1}^{\infty} \operatorname{Var}[X_j]$$

and thus,

$$P\{\omega : T(\omega) \geq \varepsilon\} \leq \frac{1}{\varepsilon^2} \sum_{j=N_1+1}^{\infty} \operatorname{Var}[X_j] .$$

Since $\sum_{j=1}^{\infty} \operatorname{Var}[X_j]$ converges, by *Theorem* (2.4.8) the right hand side of the above inequality can be made arbitrarily small by choosing N_1 sufficiently large. Thus,

$$P\{\omega : T(\omega) \geq \varepsilon\} = 0.$$

Since ε was arbitrary,

$$P\{\omega : T(\omega) > 0\} = 0.$$

Thus, $T = 0$ P.a.e. and the proof is complete.

Lemma 7.1.8

(*Toeplitz*)

Let (x_k) be a sequence of real numbers such that

$$\lim_{k\to\infty} x_k = x < \infty.$$

Let $\{a_{nk}\}$ be a set of real numbers satisfying

$$\sum_{k=1}^{n} a_{nk} = 1, \quad a_{nk} \geq 0, \quad k = 1, 2, \ldots n; \; n = 1, 2, 3, \ldots$$

and

$\lim_{n\to\infty} a_{nk} = 0$ for each k. Then $\lim_{n\to\infty} (\sum_{k=1}^{n} a_{nk} x_k) = x.$

Proof: Let $\varepsilon > 0$ be arbitrary but fixed. Since $\lim_{k\to\infty} x_k = x$, there exists a natural number N_1 such that for $n \geq N_1$, $|x_n - x| < \varepsilon$. Thus for every $n \geq N_1$ we have

$$\begin{aligned}
\left|\left(\sum_{k=1}^{n} a_{nk}x_k\right) - x\right| &= \left|\sum_{k=1}^{n} a_{nk}(x_k - x)\right| \\
&\leq \sum_{k=1}^{N_1} a_{nk}|x_k - x| \\
&\quad + \sum_{k=N_1+1}^{n} a_{nk}|x_k - x| \\
&\leq \sum_{k=1}^{N_1} a_{nk}|x_k - x| + \sum_{k=N_1+1}^{n} a_{nk}\,\varepsilon \\
&\leq \sum_{k=1}^{N_1} a_{nk}|x_k - x| + \varepsilon.
\end{aligned}$$

Hence,

$$\lim_{n\to\infty} \left| \left(\sum_{k=1}^{n} a_{nk} x_k \right) - x \right| \leq \lim_{n\to\infty} \sum_{k=1}^{N_1} a_{nk} |x_k - x| + \varepsilon.$$

However, since N_1 is fixed and $\lim_{n\to\infty} a_{nk} = 0$, we may conclude that

$$\lim_{n\to\infty} \left| \left(\sum_{k=1}^{n} a_{nk} x_k \right) - x \right| \leq \varepsilon.$$

Since $\varepsilon > 0$ was arbitrary, this implies that

$$\lim_{n\to\infty} \left| \left(\sum_{k=1}^{n} a_{nk} x_k \right) - x \right| = 0$$

as was desired.

Lemma 7.1.9

(*Kronecker's Lemma*)

Let $\{x_k\}$ be a sequence of real numbers and let $\{a_k\}$ be a sequence of positive real numbers such that $\lim_{k\to\infty} a_k = \infty$. If

$$\sum_{k=1}^{\infty} \frac{x_k}{a_k}$$

is finite then

$$\lim_{n\to\infty} \frac{1}{a_n} \sum_{k=1}^{n} x_k = 0.$$

Proof: For each natural number n let $b_n = \sum_{k=1}^{n} \frac{x_k}{a_k}$ and let $a_o = b_o = b_{-1} = 0$. Then $x_n = a_n(b_n - b_{n-1})$. By Abel's Partial Summation Formula, *Theorem* (2.4.9), we have

$$\frac{1}{a_n}\sum_{k=1}^{n} x_k = \frac{1}{a_n}\sum_{k=1}^{n} a_k(b_k - b_{k-1})$$

$$= \frac{1}{a_n}\left\{\sum_{k=1}^{n-1}\left(\sum_{j=1}^{k}(b_j - b_{j-1})\right)(a_k - a_{k+1}) + \sum_{j=1}^{n}(b_j - b_{j-1})a_n\right\}$$

$$= \frac{1}{a_n}\sum_{k=1}^{n-1} b_k(a_k - a_{k+1}) + b_n$$

$$= b_n - \frac{1}{a_n}\sum_{k=0}^{n-1} b_k(a_{k+1} - a_k).$$

For each n, we have

$$\frac{1}{a_n}\sum_{k=0}^{n-1}(a_{k+1} - a_k) = 1.$$

Note also that for fixed k,

$$\lim_{n\to\infty}\frac{1}{a_n}(a_{k+1} - a_k) = 0.$$

Thus, by Toeplitz' *Lemma* (7.1.8), we have

$$\lim_{n\to\infty}\frac{1}{a_n}\sum_{k=1}^{n} x_k = \lim_{n\to\infty}\left\{b_n - \frac{1}{a_n}\sum_{k=0}^{n-1} b_k(a_{k+1} - a_k)\right\}$$

$$= \sum_{k=1}^{\infty}\frac{x_k}{a_k} - \sum_{k=1}^{\infty}\frac{x_k}{a_k}$$

$$= 0.$$

Corollary 7.1.10

If

$$\sum_{i=1}^{\infty} \frac{a_i}{i}$$

converges, then

$$\lim_{n\to\infty} \frac{1}{n} \sum_{i=1}^{n} a_i = 0.$$

Theorem 7.1.11

Let F be the distribution function for a random variable X. If

$$\int_{-\infty}^{\infty} |x| dF(x) < \infty$$

then

$$\sum_{n=1}^{\infty} \frac{1}{n^2} \int_{-n}^{n} x^2 dF(x) < \infty$$

Proof: Let

$$a_1 = \int_0^1 x\, dF(x) + \int_{-1}^{0} |x| dF(x)$$

and

$$a_{n+1} = \int_n^{n+1} x\; dF(x) + \int_{-(n+1)}^{-n} |x| dF(x).$$

Note that

$$a_n \geq 0$$

and

$$\sum_{n=1}^{\infty} a_n = \int_{-\infty}^{\infty} |x|dF(x) < \infty.$$

Also note that

$$\int_{-(n+1)}^{-n} x^2 dF(x) + \int_{n}^{n+1} x^2 dF(x) \leq \int_{-(n+1)}^{-n} (n+1)|x|dF(x)$$

$$+ \int_{n}^{n+1} (n+1)xdF(x)$$

$$= (n+1)a_{n+1}$$

so that

$$\int_{-n}^{n} x^2 dF(x) = \sum_{k=1}^{n} [\int_{-k}^{-k+1} x^2 dF(x) + \int_{k-1}^{k} x^2 dF(x)]$$

$$\leq \sum_{k=1}^{n} ka_k .$$

Hence, we have

(1) $$\sum_{n=1}^{\infty} \frac{1}{n^2} \int_{-n}^{n} x^2 dF(x) \leq \sum_{n=1}^{\infty} \frac{1}{n^2} \sum_{k=1}^{n} k\, a_k.$$

Since the series on the right is non-negative we may rearrange the terms by use of the following:

$$\sum_{n=1}^{N_1} \frac{1}{n^2} \sum_{k=1}^{n} ka_k = \frac{1}{1^2} a_1 + \frac{1}{2^2}(a_1 + 2a_2)$$

$$+ \frac{1}{3^2}(a_1 + 2a_2 + 3a_3) + \cdots$$

$$+ \frac{1}{N_1^2}(a_1 + 2a_2 + 3a_3 + \cdots\cdots + N_1 a_{N_1})$$

$$= a_1\left[1 + \frac{1}{2^2} + \frac{1}{3^2} + \cdots + \frac{1}{N_1^2}\right] + 2a_2\left[\frac{1}{2^2} + \frac{1}{3^2} + \cdots + \frac{1}{N_1^2}\right]$$

$$+ \cdots + N_1 a_{N_1}\left[\frac{1}{N_1^2}\right]$$

$$= a_1 \sum_{n=1}^{N_1} \frac{1}{n^2} + 2a_2 \sum_{n=2}^{N_1} \frac{1}{n^2} + \cdots + N_1 a_{N_1} \sum_{n=N_1}^{N_1} \frac{1}{n^2}$$

$$= \sum_{k=1}^{N_1} k a_k \sum_{n=k}^{N_1} \frac{1}{n^2} .$$

Thus, (1) may be written as

$$(2) \qquad \sum_{n=1}^{\infty} \frac{1}{n^2} \int_{-n}^{n} x^2 dF(x) \le \sum_{n=1}^{\infty} \frac{1}{n^2} \sum_{k=1}^{n} k a_k$$

$$= \sum_{k=1}^{\infty} k a_k \sum_{n=k}^{\infty} \frac{1}{n^2} .$$

Ignoring the first term of the last series we have

$$\sum_{k=2}^{\infty} k a_k \sum_{n=k}^{\infty} \frac{1}{n^2} \le \sum_{k=2}^{\infty} k a_k \int_{k-1}^{\infty} \frac{1}{x^2}\, dx$$

$$= \sum_{k=2}^{\infty} k a_k \left[\frac{1}{k-1}\right]$$

$$= \sum_{k=2}^{\infty} \left[1 + \frac{1}{k-1}\right] a_k$$

$$= \sum_{k=2}^{\infty} a_k + \sum_{k=2}^{\infty} \frac{1}{k-1} a_k .$$

Since

$$\sum_{k=1}^{\infty} a_k = \int_{-\infty}^{\infty} |x| dF(x) < \infty$$

by assumption, each of the above series converges and hence the series on the right in (2) converges. Thus

$$\sum_{n=1}^{\infty} \frac{1}{n^2} \int_{-n}^{n} x^2 dF(x) < \infty$$

as was desired.

Theorem 7.1.12

Let $X_1, X_2, X_3, \ldots$ be a sequence of independent random variables with finite means $E[X_n] = \mu_n$ and finite variances $Var[X_n]$. If

$$\sum_{n=1}^{\infty} \frac{Var[X_n]}{n^2} < \infty$$

then

$$\lim_{n\to\infty} \left[\sum_{i=1}^{n} \frac{X_i}{n} - \sum_{i=1}^{n} \frac{\mu_i}{n} \right] = 0$$

P.a.e.

Proof: Define the random variables Y_k $k = 1, 2, \ldots$ by

$$Y_k = \frac{1}{k} X_k. \quad \text{Thus} \quad Var\ Y_k = \frac{1}{k^2} Var\ X_k$$

so that

$$\sum_{k=1}^{\infty} Var\ Y_k = \sum_{k=1}^{\infty} \frac{1}{k^2} Var\ X_k < \infty.$$

By (7.1.7), $\sum_{k=1}^{\infty} [\, Y_k - E[Y_k]] = \sum_{k=1}^{\infty} [\, \frac{X_k - E[X_k]}{k}]$

converges P.a.e. Thus by *Corollary* (7.1.10)

$$\lim_{n\to\infty} \frac{1}{n} \sum_{k=1}^{n} [X_k - E[X_k]] = 0 \quad \text{P.a.e.}$$

That is,

$$\lim_{n\to\infty} \left[\sum_{i=1}^{n} \frac{X_i}{n} - \sum_{i=1}^{n} \frac{\mu_i}{n} \right] = 0 \quad \text{P.a.e.}$$

Theorem 7.1.13

Let X be a random variable. $E[|X|] < \infty$ if and only if

$$\sum_{k=1}^{\infty} P\{\omega : |X(\omega)| \geq k\} < \infty .$$

Proof: Let

$$A_k = \{\omega : |X(\omega)| \geq k\}$$

and

$$B_k = \{\omega : k+1 > |X(\omega)| \geq k\} \text{ for } k = 0, 1, 2, 3, \ldots .$$

Assume $E[|X|] < \infty$. Note that $\{B_k : k = 0, 1, 2, \ldots\}$ is a partition of Ω.

$$\sum_{k=0}^{\infty} P[A_k] = P[B_0] + \sum_{k=1}^{\infty} (k+1)\, P[B_k]$$

$$= P[B_0] + \sum_{k=1}^{\infty} kP[B_k] + \sum_{k=1}^{\infty} P[B_k]$$

$$= P[B_o] + \sum_{k=1}^{\infty} k\, P[B_k] + 1 - P[B_o]$$

$$= 1 + \sum_{k=1}^{\infty} k\, P[B_k]$$

$$= 1 + \sum_{k=1}^{\infty} k \int_{B_k} dP, \quad \textit{Theorem } (4.2.17)$$

$$= 1 + \sum_{k=1}^{\infty} \int_{B_k} k\, dP$$

$$\leq 1 + \sum_{k=1}^{\infty} \int_{B_k} |X|\, dP$$

$$\leq 1 + \sum_{k=0}^{\infty} \int_{B_k} |X|\, dP, \quad \textit{Theorem } (4.3.10)$$

$$= 1 + \int |X|\, dP$$

$$= 1 + E[|X|] < \infty .$$

Conversely, assume that $\sum_{k=1}^{\infty} P[A_k] < \infty$.

$\int |X|\, dP = \sum_{k=0}^{\infty} \int_{B_k} |X|\, dP$ provided of course that the series on the right can be shown to converge.

Now $\sum_{k=0}^{\infty} \int_{B_k} |X|\, dP = \int_{B_o} |X|\, dP + \sum_{k=1}^{\infty} \int_{B_k} |X|\, dP$

$$\leq \int_{B_o} dP + \sum_{k=1}^{\infty} \int_{B_k} (k+1)\, dP$$

$$= P[B_o] + \sum_{k=1}^{\infty} (k+1) P[B_k]$$

$$= P[B_0] + \sum_{k=1}^{\infty} kP[B_k] + \sum_{k=1}^{\infty} P[B_k]$$

$$= 1 + \sum_{k=1}^{\infty} kP[B_k]$$

$$= P[A_0] + \sum_{k=1}^{\infty} P[A_k]$$

$$< \infty .$$

Thus,

$$E[|X|] < \infty.$$

Theorem 7.1.14

(Kolmogorov's Strong Law of Large Numbers)

Let $X_1, X_2, \ldots,$ be a sequence of independent and identically distributed random variables. Let $S_n = \sum_{k=1}^{n} X_k$. Then $(\frac{S_n}{n})$ converges P.a.e. if and only if $E[X_1]$ exists. If either condition holds, then

$$\lim_{n\to\infty} \left(\frac{S_n}{n} \right) = E[X_1] \text{ , P.a.e.}$$

Proof: Assume that $(\frac{S_n}{n})$ converges P.a.e. Then

$$\left(\frac{X_n}{n}\right) = \left(\frac{S_n}{n} - \left(\frac{n-1}{n}\right)\left(\frac{S_{n-1}}{n-1}\right)\right)$$

converges to 0, P.a.e.

Thus,

$$P\{\omega : \left|\frac{X_n(\omega)}{n}\right| \geq 1 \text{ infinitely often}\} = 0$$

which implies that

$$P\{\omega : |X_n(\omega)| \geq n \text{ infinitely often}\} = 0.$$

Now $X_1, X_2, X_3, \ldots$ are independent so that the events

$$\{\omega : |X_n(\omega)| \geq n\}$$

are also independent. Thus by the *Borel-Cantelli Lemma* (3.2.13) (contrapositive of ii) we have

$$\sum_{n=1}^{\infty} P\{\omega : |X_n(\omega)| \geq n\} < \infty .$$

However, $X_1, X_2, X_3, \ldots$ are identically distributed, so that

$$P\{\omega : |X_n(\omega)| \geq n\} = P\{\omega : |X_1(\omega)| \geq n\}.$$

Thus,

$$\sum_{n=1}^{\infty} P\{\omega : |X_1(\omega)| \geq n\} < \infty$$

and by *Theorem* (7.1.13) $E[|X_1|]$ exists. By *Theorem* (4.3.2), $E[X_1]$ exists also. Now assume that $E[X_1]$ exists. We shall show that

$$\lim_{n\to\infty} \left(\frac{S_n}{n}\right) = E[X_1] \qquad \text{P.a.e.}$$

Without loss of generality we may assume that $E[X_1] = 0$. If $E[X_1] \neq 0$, we may define $Y_i = X_i - E[X_1]$ so that

$E[Y_i] = 0$. If the theorem holds in this case, then

$$\lim_{n\to\infty} \left(\frac{1}{n} \sum_{i=1}^{n} Y_i\right) = 0.$$

But

$$\lim_{n\to\infty} \left(\frac{1}{n} \sum_{i=1}^{n} Y_i\right) = \lim_{n\to\infty} \left(\frac{1}{n} \sum_{i=1}^{n} X_i - E[X_1]\right)$$

so that

$$\lim_{n\to\infty} \left(\frac{1}{n} \sum_{i=1}^{n} X_i\right) = E[X_1]$$

and the theorem holds in general. Define random variables Y_n and Z_n by

$$Y_n = \begin{cases} X_n & \text{if} \quad |X_n| < n \\ 0 & \text{if} \quad |X_n| \geq n \end{cases}$$

and

$$Z_n = X_n - Y_n .$$

Note that

$$\begin{aligned} P\{\omega : Z_n(\omega) \neq 0\} &= P\{\omega : |X_n(\omega)| \geq n\} \\ &= P\{\omega : |X_1(\omega)| \geq n\} . \end{aligned}$$

Since

$$E[|X_1|] < \infty,$$

by *Theorem* (7.1.13),

$$\sum_{n=1}^{\infty} P\{\omega : |X_n(\omega)| \geq n\} = \sum_{n=1}^{\infty} P\{\omega : Z_n(\omega) \neq 0\} < \infty.$$

By the *Borel Cantelli Lemma*, (3.1.13),

$$P\{\omega : Z_n(\omega) \neq 0 \text{ for infinitely many } n\} = 0.$$

Thus,

$$\lim_{n\to\infty} Z_n = 0 \quad \text{P.a.e.}$$

so that

(1) $$\lim_{n\to\infty} \frac{Z_1 + Z_2 + \cdots + Z_n}{n} = 0$$

by the *Toeplitz Lemma* (7.1.8). We now consider Y_n. Due to the truncation, $\mathrm{Var}[Y_n]$ exists. Furthermore,

$$\begin{aligned}\mathrm{Var}[Y_n] &= E[(Y_n - E[Y_n])^2] \\ &= E[Y_n^2] - (E[Y_n])^2 \\ &\leq E[Y_n^2] \\ &= \int_{-n}^{n} x_1^2 \, dF(x_1) < \infty.\end{aligned}$$

By *Theorem* (7.1.11), we have

$$\sum_{n=1}^{\infty} \frac{1}{n^2} \mathrm{Var}[Y_n] \leq \sum_{n=1}^{\infty} \frac{1}{n^2} \int_{-n}^{n} x_1^2 \, dF(x_1) < \infty.$$

(2) By *Theorem* (7.1.12)

$$\lim_{n\to\infty}\left[\sum_{i=1}^{n}\frac{Y_i}{n}-\sum_{i=1}^{n}\frac{E[Y_i]}{n}\right]=0\ ,\qquad \text{P.a.e.}$$

(3) Note that since by assumption, $E[X_n] = 0$ we have that for each n, $E[Y_n] = 0$. Also

$$\left(\frac{S_n}{n}\right)=\sum_{i=1}^{n}\frac{Y_i}{n}+\sum_{i=1}^{n}\frac{Z_i}{n}$$

$$=\sum_{i=1}^{n}\frac{Y_i}{n}-\sum_{i=1}^{n}\frac{E[Y_i]}{n}+\sum_{i=1}^{n}\frac{E[Y_i]}{n}+\sum_{i=1}^{n}\frac{Z_i}{n}\ .$$

Thus,

$$\lim_{n\to\infty}\left(\frac{S_n}{n}\right)=\lim_{n\to\infty}\left(\sum_{i=1}^{n}\frac{Y_i}{n}-\sum_{i=1}^{n}\frac{E[Y_i]}{n}\right)$$

$$+\lim_{n\to\infty}\sum_{i=1}^{n}\frac{E[Y_i]}{n}+\lim_{n\to\infty}\sum_{i=1}^{n}\frac{Z_i}{n}\ .$$

Combining (1), (2) and (3) we have

$$\lim_{n\to\infty}\left(\frac{S_n}{n}\right)=0+0+0=0=E[X_1]\ ,\qquad \text{P.a.e.}$$

and the proof is complete.

Note 7.1.15

The Kolmogorov strong law of large numbers derives its name from the fact that it asserts that the sequence $\left(\frac{S_n}{n}\right)$ converges to $E[X_1]$ P.a.e. This type of convergence is strong in the sense that it implies almost uniform convergence and also convergence in probability.

7.2 *THE CENTRAL LIMIT THEOREM*

The normal distribution is of paramount importance in both theory and applications of probability. Of particular importance is the fact that, if a random variable is represented by the mean of n *independent identically distributed random variables*, $X_1, X_2, \ldots, X_n$, then their sum, S_n, for sufficiently large n, is *approximately normally distributed*, provided that the random variable possess finite means and variances. This remarkable result is known as the (classical) *Central Limit Theorem*.

Here, we shall discuss two main versions of the Central Limit Theorem: one is when the random variables in the sequence $X_1, X_2, \ldots, X_n$ are independent and identically distributed; the other is when the random variables in the sequence are independent but not identically distributed.

*Theorem** 7.2.1

Let $X_1, X_2, \ldots, X_n, \ldots$ be a sequence of independent random variables with a common distribution having mean μ and variance σ^2. Then the variate

$$Z_n = \frac{S_n - n\mu}{\sqrt{n\sigma^2}}$$

where

$$S_n = \sum_{i=1}^{n} X_i$$

is normally distributed with mean zero and variance one as $n \to \infty$.

Note 7.2.2

When we say that Z_n is normally distributed with mean zero and variance one as $n \to \infty$ we mean that

$$\lim_{n\to\infty} P[Z_n \le z] = \frac{1}{\sqrt{2\pi}} \int_{-\infty}^{z} e^{-\frac{1}{2}t^2} dt = F(z)$$

for each $z \varepsilon \mathcal{R}$. If we let F_{Z_n} represent the distribution function for the random variable Z_n and F represent the distribution function for the normal random variable with mean 0 and variance 1, then we are simply saying that for each $z \varepsilon \mathcal{R}$

$$\lim_{n\to\infty} F_{Z_n}(z) = F(z).$$

Note that the convergence involved here is nothing more than the usual convergence of a sequence of real numbers discussed in Chapter One. This mode of convergence is what will be meant throughout this chapter.

A rigorous proof of this theorem is beyond the scope of this book and the reader is referred to M. Loève [6], B. V. Gnedenko [4], A. N. Kolmogorov [5], and Fisz [3] for more extensive treatments. However, one can give a proof of the theorem under a more restricted situation in which use of the moment generating functions of the common probability distribution is made. The argument of the proof depends upon the application of *Theorem* (4.3.32) while indicating that, as $n \to \infty$, the moment generating function of the random variable Z_n approaches the moment generating function of the normal distribution with mean 0 and variance 1.

Example 7.2.3

Let $X_1, X_2, \ldots, X_n$ be a sequence of independent variates with common distribution and $E(X_i) = \mu$ and $Var(X_i) = \sigma^2$, $i = 1, \ldots, n$. Then the variate

$$Y_n = \frac{\bar{X}_n - E(\bar{X}_n)}{\sqrt{Var(\bar{X}_n)}}$$

with $\bar{X}_n = \frac{1}{n} \sum_{i=1}^{n} X_i$ is normally distributed with mean 0 and variance 1 as $n \to \infty$.

The following result is due to Lyapuov and does not require that the random variables be identically distributed.

*Theorem** 7.2.4

Let $X_1, X_2, \ldots, X_n, \ldots$ be a sequence of independent random variables with finite means $E(X_i)$; and, for some $\xi > 0$, let

$$E[|X_r - E(X_r)|]^{2+\xi}$$

be finite. Then the variate

$$Z_n = \frac{S_n - n\mu}{\sqrt{n\sigma^2}}$$

is normally distributed with mean 0 and variance 1 as $n \to \infty$, provided

$$\lim_{n\to\infty} [Var\ (S_n)]^{-(1+\frac{\xi}{2})} \sum_{r=1}^{n} E[|X_r - E(X_r)|^{2+\xi}] = 0.$$

For n sufficiently large, we can approximate the binomial probability distribution with the normal probability law. This special case of the Central Limit Theorem is known as the *DeMoivre-Laplace Theorem* which was first formulated in 1733. This theorem is of considerable theoretical and practical importance.

Theorem 7.2.5

(*DeMoivre-Laplace*)

Let $X_1, X_2, \ldots, X_n, \ldots$ be a sequence of independent random variables, each taking the values 1 or 0 with probability p or $1 - p = q$, respectively ($p \neq 0$ or 1). Then the variate

$$Z_n = \frac{S_n - np}{\sqrt{npq}}$$

is normally distributed with mean 0 and variance 1 as $n \to \infty$.

Example 7.2.6

If $X_1, X_2, \ldots, X_n, \ldots$ is a sequence of independent identically *Poisson* distributed random variables with parameter λ, then the distribution of the variate

$$Z_n = \frac{S_n - n\lambda}{\sqrt{n\lambda}}$$

is normally distributed with mean 0 and variance 1 as $n \to \infty$.

7.3 *SUMMARY*

In this chapter several of the well known classical *inequalities* and *limit theorems* were derived. In particular the following were considered:

i) *General Tchebychef's Inequality;*
ii) *Markov Inequality;*
iii) *Tchebychef's Inequality;*
iv) *Kolmogorov's Inequality;*
v) *Weak Law of Large Numbers;*
vi) *Kolmogorov's Strong Law of Large Numbers;*
vii) *Central Limit Theorem;*
viii) *DeMoivre-Laplace Theorem.*

IMPORTANT TERMS IN CHAPTER SEVEN

Tchebychef's Inequality
Markov Inequality
Weak Law of Large Numbers
Kolmogorov's Inequality
Toeplitz Lemma
Kronecker's Lemma
Strong Law of Large Numbers
Central Limit Theorem
DeMoivre-Laplace Theorem

REFERENCES AND SUGGESTIONS FOR FURTHER READINGS

[1] **Breimann, L., Probability. Reading, Massachusetts: Addison-Wesley Publishing Company, 1968.

[2] **Chung, K. L., A Course in Probability Theory. New York: Academic Press, Second Edition, 1974.

[3] Fisz, M., Probability Theory and Mathematical Statistics. New York: John Wiley and Sons, Inc. 1963.

[4] Gnedenko, B. V., The Theory of Probability. New York: Chelsea Publishing Company, 1966.

[5] Kolmogorov, A. N., Foundations of the Theory of Probability. New York: Chelsea Publishing Company, 1936.

[6] **Loève, M., Probability Theory. New York: D. Van Nostrand Company, Inc., 1963.

[7] Tsokos, C. P., Probability Distributions: An Introduction to Probability Theory with Applications. Belmont, California: Wadsworth Publishing Company, Inc., 1972.

[8] **Tucker, H. G., A Graduate Course in Probability, New York: Academic Press, 1967.

**These books are more advanced than the approach of the present text.

SOLUTIONS TO SELECTED EXERCISES

Exercises 1.1

1. i) Let $x \in A \cap (\bigcup_{\gamma\in\Gamma} B_\gamma)$. By *Definition* (1.1.12) $x \in A$ and $x \in \bigcup_{\gamma\in\Gamma} B_\gamma$. By *Definition* (1.1.11) $x \in A$ and $x \in B_\gamma$ for some $\gamma \in \Gamma$. Hence $x \in A \cap B_\gamma$ for some $\gamma \in \Gamma$. By *Definition* (1.1.11),
$x \in \bigcup_{\gamma\in\Gamma} (A \cap B_\gamma)$. Hence $A \cap (\bigcup_{\gamma\in\Gamma} B_\gamma) \subseteq \bigcup_{\gamma\in\Gamma} (A \cap B_\gamma)$.
Reversing the argument, let $x \in \bigcup_{\gamma\in\Gamma} (A \cap B_\gamma)$. By *Definition* (1.1.11), $x \in A \cap B_\gamma$ for some $\gamma \in \Gamma$. Hence by *Definition* (1.1.12) $x \in A$ and $x \in B_\gamma$. By *Definition* (1.1.11), $x \in \bigcup_{\gamma\in\Gamma} B_\gamma$. By *Definition* (1.1.12),
$x \in A \cap (\bigcup_{\gamma\in\Gamma} B_\gamma)$. Thus $\bigcup_{\gamma\in\Gamma} (A \cap B_\gamma) \subseteq A \cap (\bigcup_{\gamma\in\Gamma} B_\gamma)$.
By *Definition* (1.1.6) $A \cap (\bigcup_{\gamma\in\Gamma} B_\gamma) = \bigcup_{\gamma\in\Gamma} (A \cap B_\gamma)$.

Part ii) is proved similarly.

2. i) Apply *Theorem* (1.1.17) with $X = U$.

ii) $(A' \cap B')' = A'' \cup B'' = A \cup B$ by *Corollary* (1.1.21) and part i).

iii) $(A' \cup B')' = A'' \cap B'' = A \cap B$ by *Corollary* (1.1.21) and part i).

iv) Assume $A \subseteq B$. Let $x \in B'$. Then $x \notin B$ and hence also $x \notin A$. Thus $x \in A'$ which implies by *Definition* (1.1.3) that $B' \subseteq A'$. To prove the converse, assume that $B' \subseteq A'$. Apply the above result to obtain $A'' \subseteq B''$. By part i) $A \subseteq B$.

3. i) By *Theorem* (1.1.32)

$$\overline{\lim_{n\to\infty}} A_n = \bigcap_{n=1}^{\infty} \bigcup_{m=n}^{\infty} A_m. \text{ Hence}$$

$$(\overline{\lim_{n\to\infty}} A_n)' = [\bigcap_{n=1}^{\infty} \bigcup_{m=n}^{\infty} A_m]'$$

$$= \bigcup_{n=1}^{\infty} (\bigcup_{m=n}^{\infty} A_m)' \quad \text{by } \textit{Corollary} \text{ (1.1.21)}$$

$$= \bigcup_{n=1}^{\infty} \bigcap_{m=n}^{\infty} A_m' \quad \text{by } \textit{Corollary} \text{ (1.1.21)}$$

$$= \underline{\lim_{n\to\infty}} A_n' \quad \text{by } \textit{Theorem} \text{ (1.1.32).}$$

ii) By *Theorem* (1.1.32)

$$\underline{\lim_{n\to\infty}} A_n = \bigcup_{n=1}^{\infty} \bigcap_{m=n}^{\infty} A_m. \text{ Hence}$$

$$(\underline{\lim_{n\to\infty}} A_n)' = (\bigcup_{n=1}^{\infty} \bigcap_{m=n}^{\infty} A_m)'$$

$$= \bigcap_{n=1}^{\infty} (\bigcap_{m=n}^{\infty} A_m)' \quad \text{by } \textit{Corollary} \text{ (1.1.21)}$$

$$= \bigcap_{n=1}^{\infty} \left(\bigcup_{m=n}^{\infty} A_m' \right) \quad \text{by } \textit{Corollary } (1.1.21)$$

$$= \overline{\lim_{n\to\infty}} A_n' \quad \text{by } \textit{Theorem } (1.1.32).$$

4. $\overline{\lim_{n\to\infty}} A_{\gamma_n} = \bigcap_{n=1}^{\infty} \bigcup_{m=n}^{\infty} A_{\gamma_m}$ by *Theorem* (1.1.32).

Since $\mathcal{C}$ is a σ algebra, for each $n = 1, 2, 3, \ldots.$ $\bigcup_{m=n}^{\infty} A_{\gamma_m}$ is an element of $\mathcal{C}$. Thus $\overline{\lim_{n\to\infty}} A_{\gamma_n}$ is the intersection of a countable collection of elements of $\mathcal{C}$. By *Note* (1.1.23),

$\overline{\lim_{n\to\infty}} A_{\gamma_n} \in \mathcal{C}$. $\underline{\lim_{n\to\infty}} A_{\gamma_n} = \bigcup_{n=1}^{\infty} \bigcap_{m=n}^{\infty} A_{\gamma_m}$

by *Theorem* (1.1.32). By *Note* (1.1.23), for each $n = 1, 2, 3, \ldots$ $\bigcap_{m=n}^{\infty} A_{\gamma_m} \in \mathcal{C}$. Hence $\underline{\lim_{n\to\infty}} A_{\gamma_n}$ is the union of a countable collection of elements of $\mathcal{C}$. Since $\mathcal{C}$ is a σ algebra, $\underline{\lim_{n\to\infty}} A_{\gamma_n} \in \mathcal{C}$.

5. $\overline{\lim_{n\to\infty}} A_n = A \cup B$ and $\underline{\lim_{n\to\infty}} A_n = A \cap B$.

6. $\overline{\lim_{n\to\infty}} A_n = \bigcup_{n=1}^{\infty} A_n = \underline{\lim_{n\to\infty}} A_n$.

An example of such a sequence is the sequence $A_1, A_2, A_3, \ldots\ldots$ such that $A_n = \{1, 2, 3, \ldots\ldots n\}$.

7. $\overline{\lim_{n\to\infty}} A_n = \bigcap_{n=1}^{\infty} A_n = \underline{\lim_{n\to\infty}} A_n$. An example of such a sequence is the sequence $A_1, A_2, A_3, \ldots\ldots$ where

$A_1 = N - \{1\}$

$A_2 = N - \{1,2\}$

$A_3 = N - \{1,2,3\}$

$\cdot$

$\cdot$

$\cdot$

$A_n = N - \{1,2,3, \ldots n\}$

Exercises 1.2

1. See *Example* (1.2.2), ii).
2. See *Example* (1.1.24), iii). Let $U_x = \{x\}$. Consider $\{U_x : x \in (0, 1)\}$. $U_x \in \mathcal{C}$ for each x but $\bigcup_{x\in(0,1)} U_x = (0,1)$ is neither countable nor the complement of a countable set. Thus $\bigcup_{x\in(0,1)} U_x \notin \mathcal{C}$ which implies that $\mathcal{C}$ is NOT a topology.
3. See *Example* (1.1.24), ii).
4. i) Let $x \in A$. $f(x) \in f(A)$ implying that x is mapped by f into $f(A)$. That is, $x \in f^{-1}(f(A))$;

   ii) Consider $X = \{1, -1, 2, -2\}$ and $Y = N$. Let f be defined on X by $f(x) = x^2$. $f : X \to Y$. Let $A = \{1, 2\}$. $f(A) = \{1, 4\}$. $f^{-1}(f(A)) = X \neq A$;

   iii) Let $y \in f(f^{-1}(B))$. There exists an element $x \in f^{-1}(B)$ such that $y = f(x)$. However $x \in f^{-1}(B)$ implies that $f(x) \in B$. Hence we have $y \in B$;

iv) Let B be a subset of the range of f. Let $y \in B$. There exists an element $x \in X$ such that $y = f(x)$. Hence $x \in f^{-1}(B)$ implying that $y \in f(f^{-1}(B))$. Thus we have $B \subseteq f(f^{-1}(B))$ which together with the preceding result and *Definition* (1.1.6) implies that $B = f(f^{-1}(B))$.

5. It is obvious that $\phi \in \tau_1$ and $X \in \tau_1$. Let U and V be elements of τ_1. If $U \cap V = \phi$ then condition ii) of *Definition* (1.2.1) is trivially satisfied. Assume $U \cap V \neq \phi$. Let $p \in U \cap V$. Thus $p \in U$ and $p \in V$. By definition, there exists an interval $[a_1, b_1)$, $a_1 < b_1$ such that $p \in [a_1, b_1)$ and $[a_1, b_1) \subseteq U$. Similarly there exists an interval $[a_2, b_2)$, $a_2 < b_2$ such that $p \in [a_2, b_2)$ and $[a_2, b_2) \subseteq V$. Consider $[a_1, b_1) \cap [a_2, b_2) = [c, d)$, $c < d$. $p \in [c, d)$ and $[c, d) \subseteq U \cap V$. Thus $U \cap V \in \tau_1$. Let $\{U_\gamma : \gamma \in \Gamma\} \subseteq \tau_1$. Let $p \in \bigcup_{\gamma \in \Gamma} U_\gamma$. $p \in U_{\gamma_0}$ for some $\gamma_0 \in \Gamma$. Hence there exists an interval $[a, b)$, $a < b$ such that $p \in [a, b)$ and $[a, b) \subseteq U_{\gamma_0}$. Thus also $[a, b) \subseteq \bigcup_{\gamma \in \Gamma} U_\gamma$ which implies that $\bigcup_{\gamma \in \Gamma} U_\gamma \in \tau_1$.

6. The relationships indicated are in most instances obvious. The main result to use in attempting to verify those instances where continuity fails is *Theorem* (1.2.16).

Exercises 2.1

1. $\gamma(\phi) = \lambda(A \cap \phi) = \lambda(\phi) = 0$ and hence condition i) of *Definition* (2.1.7) is easily satisfied. Since λ is a measure and γ is defined in terms of λ, γ is obviously non-negative. Let (A_n) be a sequence of elements of $\mathcal{C}$ satisfying the hypothesis of condition iii) of *Definition* (2.1.7). Then

$$\gamma\left(\bigcup_{n=1}^{\infty} A_n\right) = \lambda\left(A \cap \bigcup_{n=1}^{\infty} A_n\right) = \lambda\left(\bigcup_{n=1}^{\infty} (A \cap A_n)\right).$$

It is evident that $(A \cap A_n)$ also satisfies the hypothesis of condition iii) of *Definition* (2.1.7) and hence

$$\lambda\left(\bigcup_{n=1}^{\infty} (A \cap A_n)\right) = \sum_{n=1}^{\infty} \lambda(A \cap A_n).$$

Thus

$$\gamma\left(\bigcup_{n=1}^{\infty} A_n\right) = \sum_{n=1}^{\infty} \lambda(A \cap A_n) = \sum_{n=1}^{\infty} \gamma(A_n)$$

as was desired.

2. That $\gamma(E) \geq 0$ for each $E \in \mathcal{C}$ is evident since γ is defined as a finite sum of non-negative real numbers.

$$\gamma(\phi) = \sum_{i=1}^{n} a_i \lambda_i(\phi) = \sum_{i=1}^{n} a_i \cdot 0 = 0$$

and hence condition i) of *Definition* (2.1.7) is easily satisfied. Let (A_j) be a sequence of elements of $\mathcal{C}$ satisfying the hypothesis of condition iii) of *Definition* (2.1.7).

$$\gamma\left(\bigcup_{j=1}^{\infty} A_j\right) = \sum_{i=1}^{n} a_i\lambda_i \left(\bigcup_{j=1}^{\infty} A_j\right)$$

$$= a_1\lambda_1\left(\bigcup_{j=1}^{\infty} A_j\right) + a_2\lambda_2 \left(\bigcup_{j=1}^{\infty} A_j\right) + \cdots a_n\lambda_n \left(\bigcup_{j=1}^{\infty} A_j\right)$$

$$= a_1 \sum_{j=1}^{\infty} \lambda_1(A_j) + a_2 \sum_{j=1}^{\infty} \lambda_2 (A_j) + \cdots\cdot a_n \sum_{j=1}^{\infty} \lambda_n(A_j)$$

$$= \sum_{i=1}^{n} a_i \sum_{j=1}^{\infty} \lambda_i(A_j)$$

$$= \sum_{i=1}^{n} \sum_{j=1}^{\infty} a_i\lambda_i(A_j)$$

$$= \sum_{i=1}^{n} \lim_{k\to\infty} \sum_{j=1}^{k} a_i\lambda_i(A_j)$$

$$= \lim_{k\to\infty} \sum_{i=1}^{n} \sum_{j=1}^{k} a_i\lambda_i(A_j)$$

$$= \lim_{k\to\infty} \sum_{j=1}^{k} \sum_{i=1}^{n} a_i\lambda_i(A_j)$$

$$= \sum_{j=1}^{\infty} \sum_{i=1}^{n} a_i\lambda_i(A_j)$$

$$= \sum_{j=1}^{\infty} \gamma(A_j)$$

as was to be shown.

Exercises 2.2

1. Define a sequence (f_n) of functions by

   $f_n(x) = n\, \chi_E(x) \quad x \in X$. Each function is obviously non-negative. To see that each is measurable, let $\alpha \in \mathcal{R}$ and note that for each n either $n \leq \alpha$ or $n > \alpha$. If $n \leq \alpha$, then $f_n^{-1}(-\infty,\alpha] = X \in \mathcal{C}$. If $\alpha < n$ then $f_n^{-1}(-\infty, \alpha] = \phi \in \mathcal{C}$ for negative α and $f_n^{-1}(-\infty, \alpha] = E' \in \mathcal{C}$ for non-negative α. In each case $f_n^{-1}(-\infty, \alpha] \in \mathcal{C}$ implying by *Theorem* (2.1.38) that for each n, f_n is measurable. We thus have the hypothesis of Fatou's Lemma satisfied. Note that if $x \in E'$, then $f(x) \leq 0$ and $f_n(x) = 0$ for each n implying that $\liminf f_n(x) = 0$ and we have $f(x) \leq \liminf f_n(x)$. If $x \in E$, then $f_n(x) = n$ for each n and $\liminf f_n(x) = \sup\{1, 2, 3, \ldots\} = \infty$. Thus again $f(x) \leq \liminf f_n(x)$. Note that for $\alpha \leq 0$, $\{x : \liminf f_n(x) \geq \alpha\} = X \in \mathcal{C}$ and for $\alpha > 0$ $\{x : \liminf f_n(x) \geq \alpha\} = E \in \mathcal{C}$. Thus $\liminf f_n$ is measurable. By *Theorem* (2.2.20)

$$0 \leq \int f d\lambda \leq \int \liminf f_n d\lambda.$$

Applying Fatou's Lemma we obtain

$$0 \leq \int f d\lambda \leq \int \liminf f_n d\lambda \leq \liminf \int f_n d\lambda.$$

However for each n, $\int f_n d\lambda = \int n X_E d\lambda = n \int X_E d\lambda$ by *Lemma* (2.2.10). Since X_E is a simple function we may apply *Definition* (2.2.7) to obtain

$\int X_E d\lambda = \lambda(E) = 0$. Thus

for each n, $\int f_n d\lambda = 0$ and we have that $0 \leq \int f d\lambda \leq 0$ implying that $\int f d\lambda = 0$ and the proof is complete.

2. Assume that $f \varepsilon M^+(X, \mathcal{C})$ and $F \varepsilon \mathcal{C}$. Let λ be a measure on $\mathcal{C}$ such that $\lambda(F) = 0$.

$\int_F f d\lambda = \int f \chi_F \, d\lambda$ by *Definition* (2.2.18).

Let $E = \{x : (f\chi_F)(x) > 0\}$. Let $K = \{x : f(x) > 0\}$. Since $f \varepsilon M^+(X, \mathcal{C})$, $K \varepsilon \mathcal{C}$ and we may write $E = F \cap K$. Since $\mathcal{C}$ is a σ algebra, $E \varepsilon \mathcal{C}$. Note that $E \subseteq F$. Note also that

$F = E \cup (E' \cap F) \cup \phi \cup \phi \cup \phi \cdots\cdots$. Hence

by (2.1.7), $\lambda(F) = \lambda(E) + \lambda(E' \cap F) + 0 + 0 + 0 + \cdots\cdot$. We may conclude therefore that $\lambda(E) \leq \lambda(F)$. This in turn implies that $0 \leq \lambda(E) \leq 0$ or that $\lambda(E) = 0$. By *Exercise* (2.2.1)

$\int f\chi_F d\lambda = 0$ implying that

$\int_F f d\lambda = 0$ as was desired.

3. Since $f \varepsilon M^+(X, \mathcal{C})$ for any $A \varepsilon \mathcal{C}$ we have $\gamma(A) = \int_A f d\lambda = \int f\chi_A d\lambda$. This last integral is obviously non-negative by *Definition* (2.2.16). $\gamma(\phi) = \int_\phi f d\lambda = 0$ by *Exercise* (2.2.2). To see that the countably additive property holds, let $A_1, A_2, \ldots\ldots$ be a sequence of elements of $\mathcal{C}$ such that $A_i \cap A_j = \phi$ $i \neq j$.

Let $A = \bigcup_{i=1}^{\infty} A_i$. We must show that

$$\gamma(A) = \gamma(\bigcup_{i=1}^{\infty} A_i) = \sum_{i=1}^{\infty} \gamma(A_i).$$

In terms of integrals, we must show that

$$\int_A f d\lambda = \int_{\bigcup_{i=1}^{\infty} A_i} f d\lambda = \sum_{i=1}^{\infty} \int_{A_i} f d\lambda.$$

Define a sequence (f_n) of maps by

$f_n = \sum_{i=1}^{n} f\chi_{A_i}$. Let $\alpha \varepsilon \mathcal{R}$. If

$\alpha < 0$, then $\{x : f_n(x) > \alpha\} = X \varepsilon \mathcal{C}$.

If $\alpha \geq 0$, then $\{x : f_n(x) > \alpha\} =$

$$\{x : \sum_{i=1}^{n} f\chi_{A_i}(x) > \alpha\}.$$

Since $A_i \cap A_j = \phi$, $i \neq j$ we can conclude that

$$\{x : \sum_{i=1}^{n} f\chi_{A_i}(x) > \alpha\} = \{x : x \varepsilon A_i \text{ for some } i,\ 1 \leq i \leq n \text{ and } f(x) > \alpha\}$$

$$= \bigcup_{i=1}^{n} (f^{-1}(\alpha, \infty) \cap A_i).$$

Since f is measurable $f^{-1}(\alpha, \infty) \varepsilon \mathcal{C}$ and since $\mathcal{C}$ is a σ algebra, $\bigcup_{i=1}^{n} (f^{-1}(\alpha, \infty) \cap A_i) \varepsilon \mathcal{C}$ implying that for each n, $f_n \varepsilon M^+(X, \mathcal{C})$ as desired. The fact that (f_n) is an increasing sequence is evident. If $x \notin A$, then for each n, $f_n(x) = 0 = f\chi_A(x)$ and thus $\lim_{n\to\infty} f_n(x) = \lim_{n\to\infty} 0 = 0 = f\chi_A(x)$.

If $x \in A$, then there exists a natural number N_1 such that $x \in A_{N_1}$. Thus for $n \geq N_1$,

$f_n(x) = f(x) = f\chi_A(x)$ implying that $\lim_{n\to\infty} f_n(x) = f\chi_A(x)$

as desired. We thus have satisfied the conditions of *Theorem* (2.2.26) and can conclude that

$$\gamma(A) = \int_A f d\lambda = \int f\chi_A d\lambda = \lim_{n\to\infty} \int f_n d\lambda = \lim_{n\to\infty} \int \sum_{i=1}^{n} f\chi_{A_i} d\lambda.$$

By *Lemma* (2.2.11) and simple induction we have that

$$\begin{aligned}
\lim_{n\to\infty} \int \sum_{i=1}^{n} f\chi_{A_i} d\lambda &= \lim_{n\to\infty} \sum_{i=1}^{n} \int f\chi_{A_i} d\lambda \\
&= \lim_{n\to\infty} \sum_{i=1}^{n} \int_{A_i} d\lambda \\
&= \lim_{n\to\infty} \sum_{i=1}^{n} \gamma(A_i) \\
&= \sum_{i=1}^{\infty} \gamma(A_i) \\
&= \sum_{i=1}^{\infty} \int_{A_i} f d\lambda.
\end{aligned}$$

Thus $\int_A f d\lambda = \sum_{i=1}^{\infty} \int_{A_i} f d\lambda$ as desired and the proof is complete.

Exercises 3.1

1. The fact that $P^*[E] \geq 0$ for each $E \varepsilon \mathcal{S}$ is evident from the fact that $P^*[E]$ is defined as a sum of a finite number of non-negative numbers.

$$P^*[\Omega] = \sum_{i=1}^{n} a_i \; P[\Omega] = \sum_{i=1}^{n} a_i$$

due to the fact that for each i, P_i is a probability measure and hence $P_i[\Omega] = 1$ for each i. However by hypothesis, $\sum_{i=1}^{n} a_i = 1$ and hence $P[\Omega] = 1$. Now let $E_1, E_2, \ldots\ldots$ be a sequence of events such that $E_i \cap E_j = \phi$ for $i \neq j$.

$$P^*[\bigcup_{j=1}^{\infty} E_j] = \sum_{i=1}^{n} a_i \; P_i[\bigcup_{j=1}^{\infty} E_j]$$

$$= a_1 P_1[\bigcup_{j=1}^{\infty} E_j] + a_2 P_2[\bigcup_{j=1}^{\infty} E_j] + \cdots\cdots a_n P_n[\bigcup_{j=1}^{\infty} E_j]$$

$$= a_1 \sum_{j=1}^{\infty} P_1[E_j] + a_2 \sum_{j=1}^{\infty} P_2[E_j] + \cdots\cdots a_n \sum_{j=1}^{\infty} P_n[E_j]$$

$$= \sum_{i=1}^{n} a_i \sum_{j=1}^{\infty} P_i[E_j]$$

$$= \sum_{i=1}^{n} \sum_{j=1}^{\infty} a_i P_i[E_j]$$

$$= \sum_{i=1}^{n} \lim_{k\to\infty} \sum_{j=1}^{k} a_i P_i [E_j]$$

$$= \lim_{k\to\infty} \sum_{i=1}^{n} \sum_{j=1}^{k} a_i P_i [E_j]$$

$$= \lim_{k\to\infty} \sum_{j=1}^{k} \sum_{i=1}^{n} a_i P_i [E_j]$$

$$= \sum_{j=1}^{\infty} \sum_{i=1}^{n} a_i P_i [E_j]$$

$$= \sum_{j=1}^{\infty} P^*[E_j] \quad \text{as was to be shown.}$$

2. Note first that $\sum_{n=1}^{\infty} \frac{1}{2^n} P_n[E]$ is term by term less than or equal to the convergent geometric series $\sum_{n=1}^{\infty} \frac{1}{2^n}$ and hence it also converges implying that P^* is a valid set function. It is obvious that $P^*[E]$ is non-negative since it is defined as the sum of a convergent series of non-negative real numbers. Consider $P^*[\Omega]$.

$$P^*[\Omega] = \sum_{n=1}^{\infty} \frac{1}{2^n} P_n[\Omega]$$

$$= \sum_{n=1}^{\infty} \frac{1}{2^n} = \frac{\frac{1}{2}}{1 - \frac{1}{2}} = 1.$$

Let $E_1, E_2, E_3, \ldots\ldots$ be a sequence of elements of $\mathcal{S}$ such that $E_i \cap E_j = \phi \quad i \neq j$. Then

$$P^*[\bigcup_{i=1}^{\infty} E_i] = \sum_{n=1}^{\infty} \frac{1}{2^n} P_n[\bigcup_{i=1}^{\infty} E_i]$$

$$= \sum_{n=1}^{\infty} \frac{1}{2^n} \sum_{i=1}^{\infty} P_n[E_i]$$

$$= \sum_{n=1}^{\infty} \sum_{i=1}^{\infty} \frac{1}{2^n} P_n[E_i]$$

$$= \sum_{i=1}^{\infty} \sum_{n=1}^{\infty} \frac{1}{2^n} P_n[E_i]$$

$$= \sum_{i=1}^{\infty} P^*[E_i].$$

3. Consider $(\mathcal{R}, \beta, \ell^*)$ and let $F_n = (-\infty, -n)$ for $n = 1, 2, 3, \ldots\ldots$. Then $F_n \supseteq F_{n+1}$ and $\ell^*(F_n) = \infty$ for each n but $\bigcap_{n=1}^{\infty} F_n = \phi$ so that $\ell^*(\bigcap_{n=1}^{\infty} F_n) = 0$. That the theorem holds with $\lambda(F_1) < \infty$ follows via the same proof as that given for *Theorem* (3.1.11) ii).

4. The fact that $P[E] \geq 0$ follows from the fact that f is non-negative.

$$P[\mathcal{R}] = \frac{1}{M} \int_{-\infty}^{\infty} f(x)\, dx = \frac{1}{M} \cdot M = 1.$$

Let $E_1, E_2, \ldots$ be a sequence of elements of β such that $E_i \cap E_j = \phi$, $i \neq j$.

$$P[\bigcup_{n=1}^{\infty} E_n] = \frac{1}{M} \int_{\bigcup_{n=1}^{\infty} E_n} f(x)dx$$

$$= \frac{1}{M} \sum_{n=1}^{\infty} \int_{E_n} f(x)dx$$

$$= \sum_{n=1}^{\infty} \frac{1}{M} \int_{E_n} f(x)dx$$

$$= \sum_{n=1}^{\infty} P[E_n].$$

Exercises 3.2

1. i) $P[E_1 \cup E_2 \cup \cdots E_n] = 1 - P[(E_1 \cup E_2 \cdots \cup E_n)']$

$= 1 - P[E_1' \cap E_2' \cdots \cap E_n']$

by *Corollary* (1.1.21).

However by *Theorem* (3.2.9) the collection $\{E_1', E_2', \ldots . E_n'\}$ is also a collection of mutually independent events. Hence by *Definition* (3.2.7);

$$P[E_1' \cap E_2' \cap \cdots E_n'] = \prod_{i=1}^{n} P[E_i'].$$

Thus

$$P[E_1 \cup E_2 \cdots \cup E_n] = 1 - \prod_{i=1}^{n} P[E_i'].$$

ii) The occurrence of exactly one event, say E_i, corresponds to the occurrence of E_i and the non-occurrence of each of the other events. That is, we are asked to find

$$P[E_i \cap E_1' \cap E_2' \cdots \cap E_{i-1}' \cap E_{i+1}' \cdots \cap E_n'].$$

By *Theorem* (3.2.9) the collection

$\{E_i, E_1', E_2'; \ldots. E_{i-1}', E_{i+1}', \ldots. E_n'\}$ is a collection of mutually independent events. Hence

$$P[E_i \cap E_1' \cap E_2' \cdots \cap E_{i-1}' \cap E_{i+1}' \cdots \cap E_n'] =$$

$$P[E_i] \prod_{\substack{k=1 \\ k\neq i}}^{n} P[E_k'] = P[E_i] \prod_{\substack{k=1 \\ k\neq i}}^{n} (1 - P[E_k]).$$

2. Let E_1 and E_2 be events such that $P[E_1] > 0$ and $P[E_2] > 0$. Assume that $E_1 \cap E_2 = \phi$. Then $P[E_1 \cap E_2] = P[\phi] = 0 \neq P[E_1]P[E_2]$. Hence E_1 and E_2 are not independent. To see that the converse is not necessarily true let $\Omega = \{\omega_1, \omega_2, \omega_3\}$; $\mathcal{S} = \rho_\Omega$. Define a set function P on $\mathcal{S}$ by

$P\{\omega_1\} = \frac{1}{2}$

$P\{\omega_2\} = \frac{1}{4}$

$P\{\omega_3\} = \frac{1}{4}$;

For $E \in \mathcal{S}$, $E \neq \phi$ $P[E] = \sum_{\omega \in E} P\{\omega\}$;

$P[\phi] = 0$.

$(\Omega, \mathcal{S}, P)$ is a probability space.

Let $E_1 = \{\omega_1, \omega_2\}$

$E_2 = \{\omega_1, \omega_3\}$.

It is obvious that E_1 and E_2 are not mutually

exclusive. Note that

$$P[E_1 \cap E_2] = P\{\omega_1\} = \tfrac{1}{2} \neq 3/4 \cdot 3/4 = P[E_1]P[E_2].$$

Thus E_1 and E_2 are also not independent. Thus the fact that two events with positive probability are not independent does not imply that they are mutually exclusive.

Exercises 3.3

1. Note that $P[\Omega] = 1$ and hence $\Omega \notin Z$. Since $P[\phi] = 0$, $\phi \in Z$. $\phi' = \Omega \notin Z$ and hence Z is not closed under complementation as is therefore not a σ algebra. Let $E \in Z$ and $F \in \mathcal{S}$. Note that $E \cap F \subseteq E$ and $E \cap F \in \mathcal{S}$. Thus $0 \leq P[E \cap F] \leq P[E] = 0$ by the first axiom of probability and *Theorem* (3.1.5). Thus $P[E \cap F] = 0$ implying that $E \cap F \in Z$. Now let $E_1, E_2, \ldots\ldots$ be a sequence of elements of Z.

   Note that

   $$0 \leq P[\bigcup_{n=1}^{\infty} E_n] \leq \sum_{n=1}^{\infty} P[E_n] = 0 \quad \text{by Boole's}$$

   inequality. Thus $P[\bigcup_{n=1}^{\infty} E_n] = 0$ implying that $\bigcup_{n=1}^{\infty} E_n \in Z$.

2. Let $F \in \mathcal{S}'$. Then $F = (E \cup Z_1) - Z_2$ for some $E \in \mathcal{S}$ and Z_1 and Z_2 in $Z \subseteq \mathcal{S}$. Since we are considering Ω as the universal set we may write

   $$F = (E \cup Z_1) \cap Z_2' = (E \cap Z_2') \cup (Z_1 \cap Z_2').$$

Since $\mathcal{S}$ is a σ algebra, $E \cap Z_2'$ and $Z_1 \cap Z_2'$ are elements of $\mathcal{S}$. This implies that $(E \cap Z_2') \cup (Z_1 \cap Z_2') = F \in \mathcal{S}$. Since $(Z_1 \cap Z_2') \subseteq Z_1$ we have expressed F in the desired form. To prove the converse, assume that $F = E \cup K \in \mathcal{S}$ where $E \in \mathcal{S}$ and $K \subseteq Z$ for some $Z \in \mathcal{Z}$. We may write

$$E \cup K = (E \cup Z) - (Z - (E \cup K)).$$

We need only show that $Z - (E \cup K) \in \mathcal{Z}$. Since $\mathcal{S}$ is a σ algebra,

$$Z - (E \cup K) = Z \cap (E \cup K)' \in \mathcal{S}.$$

Since $Z \cap (E \cup K)' \subseteq Z$ we have by *Theorem* (3.1.5) that $P[Z \cap (E \cup K)'] \leq P[Z] = 0$. Since probability measure is non-negative we may conclude that $P[Z \cap (E \cup K)'] = 0$ and hence that $Z \cap (E \cup K)' = Z - (E \cup K) \in \mathcal{Z}$ as was desired.

3. Let Ω = Natural Numbers = N

$\mathcal{S} = \{\phi, N, \{1, 3, 5, \ldots\}, \{2, 4, 6, \ldots\}\}$.

Let P be defined by

$P\{\phi\} = P\{1, 3, 5, \ldots\} = 0$

$P\{N\} = P\{2, 4, 6, \ldots\} = 1.$

$(\Omega, \mathcal{S}, P)$ is a probability space.

$\mathcal{Z} = \{\phi, \{1, 3, 5, \ldots\}\}$.

Consider $F = \{2, 4, 6, \ldots.\} \cup \{3\}$

$= \{2, 3, 4, 6, 8, \ldots\}$.

F is of the form $E \cup K$ where $E \in \mathcal{S}$ and $K \subseteq Z$ for $Z \in \mathcal{Z}$. However $E \cup K \notin \mathcal{S}$ and hence cannot be an element of $\mathcal{S}'$.

4. Note that $\phi = \phi \cup \phi$ and that $\phi \in \mathcal{S}$ and $\phi \subseteq Z$ for any element $Z \in \mathcal{Z}$. Hence $\phi \in \mathcal{S}'$, trivially. Now assume that $F \in \mathcal{S}'$. Then $F = (E \cup Z_1) - Z_2$ for $E \in \mathcal{S}$ and Z_1 and $Z_2 \in \mathcal{Z}$ by definition of $\mathcal{S}'$. However

$$(E \cup Z_1) - Z_2 = (E \cup Z_1) \cap Z_2' \quad \text{by } \textit{Note } (1.1.18).$$

Thus $F' = [(E \cup Z_1) \cap Z_2']'$

$$= (E \cup Z_1)' \cup Z_2 \quad \text{by DeMorgan's Law.}$$

Note that since $\mathcal{S}$ is a σ algebra and $E, Z_1 \in \mathcal{S}$ we have that $E \cup Z_1$ and $(E \cup Z_1)' \in \mathcal{S}$. Since $Z_2 \subseteq Z_2$ we have expressed F' in the form required in *Exercise* (3.3.2) and thus $\mathcal{S}'$ is closed under complementation. Now let $F_1, F_2, F_3, \ldots$ be a countable collection of elements of $\mathcal{S}'$. By *Exercise* (3.3.2) for each i

$F_i = E_i \cup K_i$ where $E_i \in \mathcal{S}$ and

$K_i \subseteq Z_i$ for $Z_i \in \mathcal{Z}$. Thus

$$\bigcup_{i=1}^{\infty} F_i = \bigcup_{i=1}^{\infty} (E_i \cup K_i) = \left(\bigcup_{i=1}^{\infty} E_i\right) \cup \left(\bigcup_{i=1}^{\infty} K_i\right).$$

Note that since $\mathcal{S}$ is a σ algebra $\bigcup_{i=1}^{\infty} E_i \in \mathcal{S}$.

By *Exercise* (3.3.1) $\bigcup_{i=1}^{\infty} Z_i \in \mathcal{Z}$ and obviously

$\bigcup_{i=1}^{\infty} K_i \subseteq \bigcup_{i=1}^{\infty} Z_i$. Hence $\bigcup_{i=1}^{\infty} F_i$ has been expressed

in the desired form.

5. (Well Defined) Assume that $F = E_1 \cup K_1 = E_2 \cup K_2$ where $E_1, E_2 \in \mathcal{S}$ and $K_1 \subseteq Z_1$, $K_2 \subseteq Z_2$ for Z_1 and Z_2 elements of $\mathcal{Z}$. Note that

$$E_1 \subseteq E_1 \cup K_1 = E_2 \cup K_2 \subseteq E_2 \cup Z_2$$

implies that $E_1 \subseteq E_2 \cup Z_2$. By *Theorem* (3.1.5) we obtain that

$$P[E_1] \leq P[E_2 \cup Z_2] \leq P[E_2] + P[Z_2] = P[E_2].$$

Similarly

$E_2 \subseteq E_2 \cup K_2 = E_1 \cup K_1 \subseteq E_1 \cup Z_1$ implies that $E_2 \subseteq E_1 \cup Z_1$. *Theorem* (3.1.5) implies that

$$P[E_2] \leq P[E_1 \cup Z_1] \leq P[E_1] + P[Z_1] = P[E_1].$$

Thus $P[E_1] = P[E_2]$ as desired.

(Probability Measure) Since P' is defined in terms of P which is a non-negative function, P' is trivially non-negative. $\Omega = \Omega \cup \phi$.
Hence $P'[\Omega] = P'[\Omega \cup \phi] = P[\Omega] = 1$.

Let $F_1, F_2, \ldots\ldots$ be a sequence of elements of $\mathcal{S}'$ such that $F_i \cap F_j = \phi$ $i \neq j$. Let $F_i = E_i \cup K_i$ where for each i $E_i \in \mathcal{S}$ and $K_i \subseteq Z_i$ for $Z_i \in \mathcal{Z}$.

Then

$$P'[\bigcup_{i=1}^{\infty} F_i] = P'[\bigcup_{i=1}^{\infty} (E_i \cup K_i)]$$

$$= P'[(\bigcup_{i=1}^{\infty} E_i) \cup (\bigcup_{i=1}^{\infty} K_i)].$$

Now since for each i $K_i \subseteq Z_i$, $\bigcup_{i=1}^{\infty} K_i \subseteq \bigcup_{i=1}^{\infty} Z_i$.

Also since $\mathcal{S}$ is a σ algebra,

$\bigcup_{i=1}^{\infty} E_i \in \mathcal{S}$. Thus by definition of P',

$$P'[(\bigcup_{i=1}^{\infty} E_i) \cup (\bigcup_{i=1}^{\infty} K_i)] = P[\bigcup_{i=1}^{\infty} E_i].$$

Since P is a probability measure on $\mathcal{S}$,

$P[\bigcup_{i=1}^{\infty} E_i] = \sum_{i=1}^{\infty} P[E_i]$. However by definition of P',

$$\sum_{i=1}^{\infty} P[E_i] = \sum_{i=1}^{\infty} P'[E_i \cup K_i] = \sum_{i=1}^{\infty} P'[F_i].$$

Combining these steps we obtain that

$P'[\bigcup_{i=1}^{\infty} F_i] = \sum_{i=1}^{\infty} P'[F_i]$ as was desired.

6. Let $F \in \mathcal{S}'$, $P'[F] = 0$. Assume that $A \subseteq F$. Now since $F \in \mathcal{S}'$, F can be written in the form

$F = E \cup K$ for

$E \in \mathcal{S}$ and $K \subseteq Z$ for $Z \in \mathcal{Z}$.
Since $P'[F] = P'[E \cup K] = P[E] = 0$, we can conclude also that $P[E \cup Z] = 0$ implying that $E \cup Z \in \mathcal{Z}$. Now since $A \subseteq F = E \cup K \subseteq E \cup Z$, we have that A is a subset of an element of $\mathcal{Z}$. To see that $A \in \mathcal{S}'$ note that A can be put into the form of *Exercise* (3.3.2) by writing

$$A = \phi \cup A.$$

Thus $\mathcal{S}'$ contains all subsets of sets of P' measure 0. Note that by definition of P',
$P'[A] = P'[\phi \cup A] = P'[\phi] = 0$.

Exercises 4.1

1. i) $|\ |$ is continuous on $\mathcal{R}$. Hence by *Corollary* (4.1.17), $|X|$ is a random variable;

   ii) $[\ ]$ is monotonic (increasing) on $\mathcal{R}$. Hence by *Theorem* (4.1.19), $[X]$ is a random variable;

   iii) By *Theorem* (4.1.9) $X_1 + X_2 + \cdots X_n$ is a random variable. By *Corollary* (4.1.6)
   $\frac{1}{n}(X_1 + X_2 + \cdots X_n)$ is a random variable;

   iv) By *Theorem* (4.1.3) c is a random variable. By *Theorem* (4.1.7) for each i, $X_i - c$ is a random variable. By *Theorem* (4.1.12) for each i $(X_i - c)^2$ is a random variable. By *Theorem* (4.1.9), $\sum_{i=1}^{n} (X_i - c)^2$ is a random variable.
   By *Corollary* (4.1.6), $\sum_{i=1}^{n} \frac{(X_i - c)^2}{n - 1}$ is a random

variable for any natural number $n > 1$.

v) The function $f(z) = e^z$ is continuous on R. Hence by *Corollary* (4.1.17), $f \circ X_i = e^{X_i}$ is a random variable.

2. Assume that $f : A \to R$, $X : \Omega \to A$, f is continuous with respect to the relative topology τ_A and A is a Borel set. Let α be real and consider $(-\infty, \alpha)$. We shall show that $(f \circ X)^{-1}(-\infty, \alpha) \in \mathcal{F}$ where $(\Omega, \mathcal{F}, P)$ is the underlying probability space. This is by *Theorem* (2.1.38) sufficient to guarantee that $f \circ X$ is a random variable. Now

$(f \circ X)^{-1}(-\infty, \alpha) = X^{-1}(f^{-1}(-\infty, \alpha))$. Since f is continuous and $(-\infty, \alpha)$ is open with respect to the usual topology E on R, $f^{-1}(-\infty, \alpha) \in \tau_A$ by *Theorem* (1.2.16). Hence by *Definition* (1.2.9) there exists an element $U \in E$ such that $f^{-1}(-\infty, \alpha) = A \cap U$. By *Theorem* (2.1.41) U is a Borel set. Hence $A \cap U$ is a Borel set. Since X is a random variable, by *Definition* (4.1.1) and *Definition* (2.1.34)

$X^{-1}(A \cap U) = X^{-1}(f^{-1}(-\infty, \alpha)) \in \mathcal{F}$ and the proof is complete.

3. i) Let $f(z) = \ln z$ and $A = R_+$. The conditions of *Exercise* (4.1.2) are met.

ii) Let $f(z) = \sqrt{z}$ and $A = R_+ \cup \{0\}$. The conditions of *Exercise* (4.1.2) are met.

iii) Let $f(z) = \frac{1}{z}$ and $A = R - \{0\}$. The conditions of *Exercise* (4.1.2) are met.

Exercises 4.3

1. Let $E = \{\omega : |X(\omega)| \geq \varepsilon\}$. For $\omega \in E$, $|X|^n > \varepsilon^n$. By *Definition* (4.3.1), $E[|X|^n] = \int |X|^n \, dP$. By *Theorem* (4.3.10),

$$\int |X|^n \, dP = \int_E |X|^n \, dP + \int_{E'} |X|^n \, dP$$

$$\geq \int_E |X|^n \, dP$$

$$\geq \int_E \varepsilon^n \, dP \quad \text{by } \textit{Theorem } (2.2.20)$$

$$\geq \varepsilon^n \int_E dP$$

$$= \varepsilon^n P[E]$$

$$= \varepsilon^n P[|X| \geq \varepsilon].$$

Thus $E[|X|^n] \geq \varepsilon^n P[|X| \geq \varepsilon]$ or

$$P[|X| \geq \varepsilon] \leq \frac{E[|X|^n]}{\varepsilon^n} \quad \text{and}$$

the proof is complete.

2. Apply *Exercise* (4.3.1) with $X - \mu$ playing the role of X and $n = 2$ to obtain

$$P[|X-\mu| \geq \varepsilon] \leq \frac{E[(X-\mu)^2]}{\varepsilon^2} = \frac{\text{Var } X}{\varepsilon^2}.$$

3. By *Theorem* (4.3.25), $\text{Var } aX = E[(aX)^2] - (E[aX])^2$

$$= a^2 E[X^2] - a^2 (E[X])^2$$

$$= a^2 (E[X^2] - (E[X])^2)$$

$$= a^2 \text{ Var } X.$$

4. Let $E_i = (\frac{1}{i+1}, \frac{1}{i}]$ $i = 1, 2, 3, \ldots\ldots$

Note that $\Omega = \bigcup_{i=1}^{\infty} E_i$ and $E_j \cap E_k = \phi$ for $j \neq k$.

Define $X : \Omega \to R$ by

$X(\omega) = (-1)^i(i+1)$ for $\omega \varepsilon E_i$.

Then
$$\int_{E_i} XdP = \int_{E_i} (-1)^i(i+1)\, dP$$
$$= (-1)^i(i+1) \int_{E_i} dP$$
$$= (-1)^i(i+1)\left(\frac{1}{i} - \frac{1}{i+1}\right)$$
$$= (-1)^i \cdot \frac{1}{i}.$$

Also $\sum_{i=1}^{\infty} \int_{E_i} XdP = \sum_{i=1}^{\infty} (-1)^i \frac{1}{i}$ converges since this is an alternating harmonic series. Thus the given conditions are satisfied. To show that $\int XdP = E[X]$ fails to exist, we shall show using *Exercise* (2.2.3) that $E[|X|]$ fails to exist and apply *Theorem* (4.3.2).

$$E[|X|] = \int_{(0,1]} |X|dP = \int_{\bigcup_{i=1}^{\infty} E_i} |X|dP = \sum_{i=1}^{\infty} \int_{E_i} |X|dP$$

by *Exercise* (2.2.3).

However
$$\sum_{i=1}^{\infty} \int_{E_i} |X|dP = \sum_{i=1}^{\infty} \int_{(\frac{1}{i+1}, \frac{1}{i}]} (i+1)dP$$
$$= \sum_{i=1}^{\infty} (i+1)\left(\frac{1}{i} - \frac{1}{i+1}\right)$$
$$= \sum_{i=1}^{\infty} \frac{1}{i}.$$

The latter series is the divergent harmonic series and hence $E[|X|]$ does not exist. By *Theorem* (4.3.2), $E[X]$ does NOT exist.

5. By definition $X(\omega) = (-1)^i (i+1)$ for $\omega \in E_i$ where $\{E_i\}_{i \in N}$ is a partition of $(0, 1]$. Thus the range of X is the set $\{-2, 3, -4, 5, -6, ...\}$ which is countable. Furthermore

$$f(x_i) = P[X = x_i] = P[\{\omega : X(\omega) = x_i\}] = P[E_i]$$

$$= \int_{E_i} dP$$

$$= \frac{1}{i} - \frac{1}{i+1}$$

$$= \frac{1}{i(i+1)}.$$

Thus the density for X is given by

$$f(x_i) = \begin{cases} \dfrac{1}{i(i+1)} & x_i = (-1)^i (i+1) \quad i = 1, 2, 3, ... \\ 0 & \text{elsewhere.} \end{cases}$$

As was seen in *Exercise* (4.3.4),

$$\sum_{i=1}^{\infty} (-1)^i (i+1) \frac{1}{i(i+1)} = \sum_{i=1}^{\infty} \frac{(-1)^i}{i} \quad \text{converges.}$$

Hence $\sum_{\text{all } x} xf(x)$ is finite. However,

$$\sum_{\text{all } x} |x| f(x) = \sum_{i=1}^{\infty} |(-1)^i (i+1)| \frac{1}{i(i+1)} = \sum_{i=1}^{\infty} \frac{1}{i}$$

is the divergent harmonic series. Hence the conditions of *Theorem* (4.3.12) are NOT met and $E[X]$ does NOT exist.

6. Assume that X is continuous with density f and that $E[X]$ exists. By *Theorem* (4.3.18)

$$\int_{-\infty}^{\infty} |x| f(x)\,dx \quad \text{is finite.}$$

By *Definition* (2.4.2), $\int_{-\infty}^{0} |x| f(x)\,dx$ and $\int_{0}^{\infty} |x| f(x)\,dx$ are finite. Now

$$\begin{aligned}\int_{-\infty}^{\infty} xf(x)\,dx &= \int_{-\infty}^{0} xf(x)\,dx + \int_{0}^{\infty} xf(x)\,dx \\ &= -\int_{-\infty}^{0} -x\, f(x)\,dx + \int_{0}^{\infty} xf(x)\,dx \\ &= -\int_{-\infty}^{0} |x|\, f(x)\,dx + \int_{0}^{\infty} |x|\, f(x)\,dx.\end{aligned}$$

Since each of the latter integrals is finite, their sum is finite and

$$\int_{-\infty}^{\infty} xf(x)\,dx \quad \text{exists.}$$

7. The inequality is trivially true. By *Theorem* (4.3.6) and *Corollary* (4.3.5), $E[1 + |X^n|] = 1 + E[|X^n|]$. By *Theorem* (4.3.2), $E[|X^n|]$ exists and hence $1 + E[|X^n|]$ is finite. By *Theorem* (4.3.3), $E[X^n]$ exists.

8.

$$m_X(\theta) = \begin{cases} \frac{1}{\theta(b-a)}\,[e^{\theta b} - e^{\theta a}] & \theta \neq 0 \\ 1 & \theta = 0 \end{cases}$$

$$\frac{dm_X(0)}{d\theta} = \lim_{\theta \to 0} \frac{\left(\frac{1}{\theta(b-a)}\,[e^{\theta b} - e^{\theta a}] \quad -1\right)}{\theta}$$

$$= \lim_{\theta \to 0} \frac{e^{\theta b} - e^{\theta a} - \theta(b-a)}{\theta^2 (b-a)}$$

$$= \lim_{\theta \to 0} \frac{be^{\theta b} - ae^{\theta a} - (b-a)}{2\theta(b-a)}$$

$$= \lim_{\theta \to 0} \frac{b^2 e^{\theta b} - a^2 e^{\theta a}}{2(b-a)}$$

$$= \frac{b^2 - a^2}{2(b-a)}$$

$$= \frac{b + a}{2} .$$

Exercises 5.2

1. Assume that (X_n), X, and g are as described in the exercise. Note that by *Corollary* (4.1.17) $g(X)$ and $g(X_n)$ are random variables. Let $A = \{\omega : X_n(\omega) \not\to X(\omega)\}$ and note that by *Note* (5.1.6) $A \in \mathcal{S}$ and $P[A] = 0$. Choose $\omega_o \in A'$ and note that $X_n(\omega_o) \to X(\omega_o)$. It is sufficient to show that $(g(X_n(\omega_o))) \to g(X(\omega_o))$. Choose $\varepsilon > 0$. Since g is continuous on $\mathcal{R}$ g is in particular continuous at $X(\omega_o)$. This implies that there exists a positive number $\sigma_\varepsilon(X(\omega_o))$ such that for real numbers z such that $|z - X(\omega_o)| < \sigma_\varepsilon(X(\omega_o))$ we have $|g(z) - g(X(\omega_o))| < \varepsilon$. Since $X_n(\omega_o) \to X(\omega_o)$ there exists a natural number N_1 such that $n \geq N_1$ implies $|X_n(\omega_o) - X(\omega_o)| < \sigma_\varepsilon(X(\omega_o))$. Thus for $n \geq N_1$, $|g(X_n(\omega_o)) - g(X(\omega_o))| < \varepsilon$ implying that $(g(X_n(\omega_o))) \to g(X(\omega_o))$.

2. Assume that g is not continuous at some point x_0. Then there exists an $\varepsilon^* > 0$ such that given any open interval D about x_0 there exists an $x \in D$ such that $|g(x) - g(x_0)| \geq \varepsilon^*$. Let us define a sequence (D_n) of intervals about x_0 by

$$D_n = \{x : |x-x_0| < \frac{1}{n}, n = 1, 2, 3, \ldots\}.$$

Choose a sequence of points (x_n) such that $x_n \in D_n$ and $|g(x_n) - g(x_0)| \geq \varepsilon^*$. Define a sequence (X_n) of random variables by $X_n(\omega) \equiv x_n$ and define a random variable X by $X(\omega) \equiv x_0$. We claim that (X_n) converges to X everywhere and hence by *Theorem* (5.1.7) (X_n) also converges to X with probability one. To see this let $\omega \in \Omega$ and choose $\varepsilon > 0$. There exists a natural number N_1 such that for $n \geq N_1$, $\frac{1}{n} \leq \frac{1}{N_1} < \varepsilon$. Note that $|X_n(\omega) - X(\omega)| = |x_n - x_0| < \frac{1}{n}$. Thus for $n \geq N_1$, $|X_n(\omega) - X(\omega)| < \frac{1}{n} < \frac{1}{N_1} < \varepsilon$ implying that $(X_n) \to X$. We claim that $(g(X_n))$ does NOT converge to $g(X)$ for any $\omega \in \Omega$ and hence

$(g(X_n)) \underset{\text{P.a.e.}}{\not\to} g(X)$. To see this simply let $\varepsilon = \varepsilon^*$ and let $\omega \in \Omega$. Then

$g(X_n(\omega)) = g(x_n)$ and $g(X(\omega)) = g(x_0)$. For each natural number n

$$|g(X_n(\omega)) - g(X(\omega))| = |g(x_n) - g(x_0)| \geq \varepsilon^*$$

by construction. Hence $(g(X_n)) \underset{P.a.e.}{\not\to} g(X)$.

3. Assume that $(X_n) \underset{m.s.}{\to} X$. Note that by Exercise (4.3.1)

$$P[|X_n - X| \geq \alpha] \leq \frac{E[|X_n - X|^2]}{\alpha^2}$$ for each natural

number n and each $\alpha > 0$. Thus

$$\lim_{n\to\infty} P[|X_n - X| \geq \alpha] \leq \lim_{n\to\infty} \frac{E[|X_n - X|^2]}{\alpha^2}.$$

However since $(X_n) \underset{m.s.}{\to} X$,

$$\lim_{n\to\infty} \frac{E[|X_n - X|^2]}{\alpha^2} = \frac{1}{\alpha^2} \lim_{n\to\infty} E[|X_n - X|^2] = 0.$$

Thus we have

$$0 \leq \lim_{n\to\infty} P[|X_n - X| \geq \alpha] \leq 0$$ implying

that $(X_n) \underset{P}{\to} X$.

Exercises 6.2

2. (*Tchebychef's Theorem*)

Note that $E[\frac{1}{n}\sum_{i=1}^{n} X_i] = \frac{1}{n}\sum_{i=1}^{n} E[X_i]$

and that

$$\text{Var }[\frac{1}{n}\sum_{i=1}^{n} X_i] = \frac{1}{n^2}\sum_{i=1}^{n} \text{Var } X_i < \frac{1}{n^2}\, nc = \frac{c}{n}.$$

Apply Tchebychef's inequality to obtain

$$P\{\omega : |\frac{1}{n}\sum_{i=1}^{n} X_i - \frac{1}{n}\sum_{i=1}^{n} E[X_i]| < \alpha\}$$

$$\geq 1 - \frac{\text{Var}[\frac{1}{n}\sum_{i=1}^{n} X_i]}{\alpha^2}$$

$$\geq 1 - \frac{c}{n\alpha^2}.$$

Taking limits,

$$\lim_{n\to\infty} P\{\omega : |\frac{1}{n}\sum_{i=1}^{n} X_i - \frac{1}{n}\sum_{i=1}^{n} E[X_i]| < \alpha\}$$

$$\geq \lim_{n\to\infty} (1 - \frac{c}{n\alpha^2})$$

$$= 1.$$

Since each term of the above sequence is a probability and therefore has value less than or equal to 1, we have that

$$\lim_{n\to\infty} P\{\omega : |\frac{1}{n}\sum_{i=1}^{n} X_i - \frac{1}{n}\sum_{i=1}^{n} E[X_i]| < \alpha\} \leq 1.$$

By the antisymmetric property of the real numbers

$$P\{\omega : |\frac{1}{n}\sum_{i=1}^{n} X_i - \frac{1}{n}\sum_{i=1}^{n} E[X_i]| < \alpha\} = 1.$$

4. *(Poisson's Theorem)*

For each i define a random variable X_i by

$$\begin{cases} X_i = 1 & \text{if a success occurs on the } i^{th} \text{ trial} \\ X_i = 0 & \text{otherwise.} \end{cases}$$

Thus $\begin{cases} P[X_i=1] = p_i \\ P[X_i=0] = 1 - p_i. \end{cases}$

Also $\begin{cases} E[X_i] = p_i \\ Var[X_i] = p_i - p_i^2 \leq \frac{1}{4}. \end{cases}$

Define $Z_n = \sum_{i=1}^{n} X_i$. Then

$$E[\frac{Z_n}{n}] = \frac{1}{n} E[Z_n] = \frac{1}{n} E[\sum_{i=1}^{n} X_i] = \frac{1}{n}\sum_{i=1}^{n} E[X_i]$$

$$= \frac{1}{n}\sum_{i=1}^{n} p_i.$$

We have satisfied the hypothesis of Tchebychef's Theorem and may conclude that

$$P\{\omega : | \frac{Z_n}{n} - \frac{1}{n}\sum_{i=1}^{n} p_i| < \alpha\} = 1.$$

5. The proof of this theorem is analogous to the proof of Poisson's Theorem.

6. Since $E[\frac{1}{n}\sum_{i}^{n} X_i] = \frac{1}{n}\sum_{i=1}^{n} E[X_i] = \frac{1}{n} \cdot n\, a = a,$

this theorem is just a special case of Tchebychef's Theorem.

INDEX

INDEX